MOTOR CONTROL AND LEARNING

MOTOR CONTROL AND LEARNING

Edited by

Mark L. Latash

THE PENNSYLVANIA STATE UNIVERSITY

and

Francis Lestienne

UNIVERSITÉ DE CAEN BASSE-NORMANDIE, FRANCE

Springer

Library of Congress Cataloging-in-Publication Data

Motor control and learning / edited by Mark L. Latash and Francis Lestienne.
p. ; cm.
Includes bibliographical references and index.
ISBN-13: 978-0-387-25390-9 (alk. paper)
ISBN-10: 0-387-25390-4 (alk. paper)
1. Motor learning. 2. Cognition. 3. Movement, Psychology of. I. Latash, Mark L., 1953- II. Lestienne, Francis.
[DNLM: 1. Movement—physiology. 2. Learning—physiology. 3. Psychomotor Performance—physiology. WE 103 M9167 2006]
QP301.M6855 2006
612.8′11—dc22

2005051575

Printed in the United States of America.

9 8 7 6 5 4 3 2 1 SPIN 117394

springeronline.com

CONTENTS

PREFACE

The purpose of the current volume is two-fold. First, it presents a series of review papers reflecting the recent progress in the area of neural control of posture and movement (Parts I and II). Second, it focuses on issues of changes in motor patterns and neurological structures involved in their production with learning, development, and aging (Parts III and IV).

The chapters in this volume were written by speakers at the Fourth meeting "Progress in Motor Control" that took place in Caen (France) in 2003. As such, it continues the tradition of a series "Progress in Motor Control" with the first three volumes published in 1999, 2002, and 2003. The authors of the chapters were explicitly encouraged to present state-of-the-art review of particular aspects of motor control and to use their own studies as illustrations of the most important developments in the area.

As in all previous volumes, we hope that the spirit of Nikolai Bernstein can be perceived in all the chapters. Nikolai Bernstein, viewed by many as the father of contemporary motor control, was deeply involved in studies of the effects of learning on motor coordination. These activities resulted in a classical book "Dexterity and Its Development" written in the mid-nineteen-forties and published in English in 1996. Bernstein also contributed to several seminal studies in the area of motor development. He used a unique method of measuring movement kinematics, cyclogrammetry, which he had himself developed, and quantified changes in locomotor patterns happening during the first years of age.

The opening chapter of the volume written by Anatol Feldman addresses very basic issues of motor control, those related to the nature of variables that are manipulated by the central nervous system to produce natural movements. The chapter makes a strong argument in favor of parametric control of the neuromotor apparatus contrasting it with attempts to develop control schemes that try to prescribe patterns of state variables such as forces, displacements, or muscle activation patterns. Feldman briefly reviews the equilibrium-point hypothesis of motor control, which he suggested about 40 years ago, and illustrates its power in dealing with a spectrum of motor problems including such evergreen problems as the relation between posture and movement and the problem of motor redundancy.

The second chapter is co-authored by Rosenbaum, Cohen, Meulenbroek, and Vaughan. The authors address in this chapter another central issue of motor control, that of creating motor plans. In line with theorizing by David Rosenbaum and his colleagues, this chapter develops the idea of end-state comfort as an organizing criterion for the formation motor plans. The chapter also highlights the role of mental representation in motor control.

Chapter 3 focuses on issues of postural control. Bouisset, Le Bozec, and Ribreau consider an aspect of the control of vertical posture that has been typically overlooked in many earlier studies. Their chapter deals with the question of the interface between the body and its physical environment, namely adherence and friction. The authors used a particular experimental method involving the application of self-imposed postural perturbations to study a quantitative index of adherence. They also develop a biomechanical model that allows analyzing the mechanical behavior of the postural chain.

The second part of the volume unites three chapters written by leaders in the area of studies of cyclic actions. Ramesh Balasubramaniam in Chapter 4 reviews recent studies that link the ideas from the trajectory formation area to timing accuracy in repetitive movements. This chapter also offers a controversial paradigm that tries to bring together two approaches to motor control that have traditionally been viewed as incompatible, the dynamic systems approach and the information processing approach.

The fifth chapter by Dagmar Sternad focuses on a particular motor task, bouncing a ball on the tennis racket to address several basic aspects of the production of rhythmic actions such as their stability in the presence of external perturbations and spontaneous variability of the motor pattern. Sternad describes a discrete non-linear model reflecting the kinematics of the ball and the racket and their interactions during impact. The model is used to predict stable regimes of ball bouncing, which are then compared to actual performance of humans. Based on her studies, Sternad concludes that human actors sense and make use of stability properties of this task.

The last chapter in the second part of the book by Saltzman, Nam, Goldstein, and Byrd addresses issues that are in some aspects similar to those reviewed by Feldman in Chapter 1. In particular, these authors analyze skilled motor behavior in terms of state-, parameter-, and graph-dynamics. After a review of these concepts, Saltzman and his colleagues focus on the manner in which variation in dynamical graph structure can be used to explicate the temporal patterning of speech. They present simulations of speech gestural sequences using the task-dynamic model of speech production.

The five chapters of Part III review very different aspects of changes in motor patterns and neurophysiological structures associated with motor learning. Bardy, Faugloire, Fourcade and Stoffregen suggest a model of vertical posture and then describe changes that happen when human subjects are asked to learn a novel coordination among major postural joints. This chapter addresses central issues of stabilization and destabilization that accompany the process of motor skill acquisition. It also reviews similarities and differences between the processes of learning a postural coordination and a bimanual coordination.

The motor cortex and the cerebellum have traditionally been in the center of attention of motor control researchers. Recently, a number of studies have suggested that the motor cortex is not simply an executor of motor commands and that it is involved in different aspects of motor learning. Chapter 8 by Mark Hallett address the role of the motor cortex in motor learning. Mark Hallett reviews studies using brain stimulation techniques such as transcranial magnetic stimulation (TMS), traditional electroencephalographic methods (EEG), and brain imaging techniques (positron emission tomography, PET) to demonstrate plastic changes in the motor cortex during different phases of motor learning. These studies have shown, in particular, the important role of motor cortex during implicit motor learning and during the stage of consolidation.

In Chapter 9, Michael Graziano reviews his recent exciting studies of the effects of relatively long-lasting electrical stimulation of the motor cortex of primates that has been shown to induce multi-joint movements resembling common gestures in the monkey's behavior. Experiments by Graziano and his colleagues suggest that the mapping between cortex and muscles may continuously change depending on proprioceptive feedback from the limb. This "feedback remapping" may play a fundamental role in motor control, allowing motor cortex to flexibly control different aspects of movement.

The interaction between the cerebal cortex and the cerebellum is addressed in Chapter 10 by Frolov and Dufosse. These authors offer a neural network model developed based on the column organization of the cerebral cortex and the Marr-Albus-Ito theory of cerebellar learning. The model assumes synaptic plasticity in the cerebral cortex, in the cerebellar cortex, and in the cerebellar-thalamo-cortical pathway. The model demonstrates that adaptive processes that take place in different sites of the cerebral cortex and the cerebellum do not interfere but complement each other during learning of arm reaching movement, and that any linear combination of the cerebral motor commands may generate signals able to drive the cerebellar learning processes.

Issues of the role of the cerebellum in motor control and learning are also addressed in Chapter 11 by Lalonde and Strazielle. These researchers used a unique animal model of mutant mice with cerebellar atrophy. Motor performance of the mutant mice in a set of motor tasks was shown to correlate with changes in the activity of a mitochondrial enzyme, cytochrome oxidase, in the cerebellar cortex and deep nuclei. These results suggest that changes in the cerebellum are responsible for behavioral differences in the used motor tasks. They also showed that staining of this particular enzyme may be useful as a predictor of motor capacity.

Two large chapters form the last part of the volume. Cheron, Cebolla, Leurs, Bengoetxea, and Dan discuss in detail the issue of changes in intersegmental coordination during development of locomotion. The general pattern of intersegmental coordination and the stabilization of the trunk with respect to vertical are immature at the onset of unsupported walking in toddlers, but they develop in parallel very rapidly in the first few weeks of walking experience. The authors describe a dynamic recurrent neural network, which is able to reproduce lower limb kinematics in toddler locomotion by using multiple raw EMG data. In the context of motor learning such a network may be considered as a model of biological learning mechanisms underlying motor adaptation.

The final chapter deals with changes in motor coordination that accompany natural aging. Latash, Shim, Shinohara, and Zatsiorsky consider age-related changes in the hand neuromuscular apparatus and accompanying changes in both finger strength and finger coordination. They use analysis of performance in maximal and submaximal effort tasks with different degree of involvement of intrinsic and extrinsic muscle groups. These studies have suggested a disproportionate loss of force by intrinsic hand muscles, which may have important implications for multi-digit synergies. They have also shown a deficit in the ability to stabilize the total force and the total moment produced by a set of digits in both pressing and grasping tasks. The chapter discusses a possibility that some of the age-related changes may be viewed as adaptive, while other changes are more likely to interfere with the everyday hand function making it suboptimal.

This volume may be recommended to a broad range of researchers working in the areas of aging, biomechanics, development, kinesiology, motor control, neurophysiology, neuroscience, psychology, robotics and related areas. It may also be used as a supplementary reading for graduate students in these areas.

Acknowledgments

The organization of the "Progress in Motor Control IV" conference would not have been possible without the generous support of:

- Université de Caen Basse-Normandie
- Conseil Régional de Basse-Normandie
- Conseil Général du Calvados
- Mairie de Caen
- Centre National de la Recherche Scientifique (CNRS)
- UFR STAPS & Centre de Recherche en Activité Physique et Sportive (EA 2131) Université de Caen
- Maison de la Recherche en Sciences Humaines (UMS CNRS 843) et PPF ModeScos (Plan Pluri Formation ≪ Modélisation en Sciences Cognitives & Sociales) Université de Caen

We are very grateful to Prof. Francine Thullier, Prof. Pierre Denise & Christophe Bertrand who shared the load of organizing and running the meeting.

Mark L. Latash and Francis G. Lestienne

I. CONTROL OF MOVEMENT AND POSTURE

1. THE NATURE OF VOLUNTARY CONTROL OF MOTOR ACTIONS

Anatol G. Feldman

Department of physiology, University of Montreal, Canada

Abstract

Natural laws express the relationships between certain variables called state variables. Constrained by natural laws these variables cannot be specified directly by the nervous system, as illustrated by the failure of the force control theory that relies on the idea of direct programming of kinematics and muscle torques. Natural laws include parameters, some of which are not conditioned by these laws but define essential characteristics of the system's behavior under the action of these laws. This implies that the neural control of motor actions involves changes in parameters of the system. This strategy allows the nervous system to take advantage of natural laws in producing the desired motor output without actually knowing these laws or imitating them in the form of internal models. A well established form of parametric control—threshold control—is briefly reviewed with a major focus on how it helps to solve several motor problems, in particular, the problem of the relationship between posture and movement and redundancy problems in the control of multiple muscles and joints.

The Nature of Voluntary Control of Motor Actions

THE ESSENCE OF CONTROL PROCESSES IMPLIED BY NATURAL LAWS

Laws of mechanics are universal, which implies, in particular, that they are equally applied to non-living bodies like stones or biological, living organisms like human beings. Therefore, the description and analysis of movements of biological systems is primarily relies on mechanical laws. It is clear, however, that motion of living and non-living systems are fundamentally different. We usually emphasize this difference by saying that movements of the former are *controlled* whereas those of the latter are not. This statement does not tells us much about the essence of the difference since the word "controlled" is not self-explanatory and is unclear without a specific definition. Indeed, one can study control system theories in the attempt to find out a definition of the notion. Although succeeded in the description of many control principles applied to artificial machines, including robots, these theories do not go far enough to be considered physiologically feasible. Attempts to directly apply such principles to biological systems have been made in the past and are undertaken recently but proved to be unsuccessful (Ostry and Feldman, 2003).

The most recent theory of this kind is based on the idea of programming of muscle forces to produce a desired goal-directed motion. A departure point of this theory is the fact that laws of mechanics relate kinetic (forces, torques) and kinematic variables (primarily, acceleration). This point is combined with the believe that the relationships inherent in laws of mechanics are imitated by some neural structure called an internal model (Hollerbach 1982). It is further assumed that this model is used by the system to calculate and specify muscle forces according to the desired kinematic output. In other words, according this theory, control levels of the nervous system directly deal with and calculate forces (and appropriate EMG signals) required for the production of voluntary movements.

A major problem of this force control strategy is that it implies that, before the movement execution, the system plans its desired kinematic characteristics and specify appropriate muscle forces. In other words, this strategy implies a certain cause-effect relationship in movement production—that the kinematics dictates the forces generated in the system. This strategy cannot substitute laws of mechanics that imply the opposite cause-effect relation: that forces dictate changes in kinematics, rather than the other way around. The combination of the two conflicting ideas makes the force control theory inconsistent with many physiological phenomena (for review see Ostry and Feldman 2003). In particular, it failed to explain how the system produces movement without evoking resistance of posture stabilizing mechanisms to the deviation from the initial posture (for detail see also Feldman and Latash, 2005). This drawback of the theory is not diminished by its success in the explanation of the evolution of hand trajectories and velocity profiles in pointing movements during adaptation to different force fields—other theories explain the same phenomenon without running in the posture-movement problem (Gribble and Ostry, 2000).

To clarify the notion of control that may be applied to biological systems we need to consider very general characteristics of natural laws. These laws express the relationships between certain variables called state variables (SVs; e.g., forces and kinematic variables in laws of mechanics). Constrained by natural laws, SVs, cannot be specified directly by the nervous system, as illustrated by the failure of the force control theory that relies on the idea of direct programming of SVs.

In this situation, how can the nervous system control motor behavior? A general answer to this question is the following. Natural laws include parameters, some of which are not conditioned by these laws but define essential characteristics of the system's behavior under the action of the laws. Fig. 1 shows the difference between SVs and parameters in a simple physical system—a pendulum (a mass on a rope). Note that the pendulum oscillates about a position at which the system can reach a steady state when the oscillations decay. In this steady or equilibrium state, all forces are balanced. However, it is not forces (or other SVs) but the system's parameters that pre-determine where, in the force-position space, this state can be achieved. The frequency of oscillations is also defined by parameters—by the length of the rope from which the mass is suspended and by the gravitational constant. The vertical orientation about which the pendulum oscillates is also determined by another parameter—the local direction of gravity. By changing parameters, for example, the coordinates of the suspension point in a pendulum, one can transfer the oscillations to a new location in space (Fig. 1). Similarly,

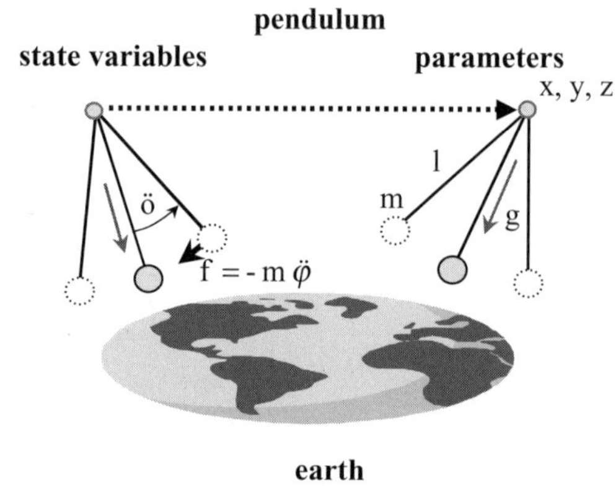

FIGURE 1. State variables (SVs), parameters, and parametric control. Related by the law of mechanics, $f = -m\,\ddot{\varphi}$, the force (f) acting on the mass of the pendulum and kinematic variables (position, φ, and its time derivatives) are SVs. The coordinates of the suspension point (x, y, z), the length (l) of the pendulum, the mass (m) and the local direction of gravity (red arrows) are parameters, i.e. quantities that can be specified independently of SVs, for example by a person who made the pendulum. The system's behavior can be controlled without direct specification of forces or other SVs, by changing parameters, for example, the coordinates of the suspension point, thus transferring the oscillations to a new location in space (dashed arrow). Frequency of oscillations can be controlled by changing parameter λ. (reproduced from Feldman 2005)

in neuromuscular systems, in order to elicit a motor action, neural control levels must change parameters that are independent of SVs. Our motor skills are thus based on the ability of the brain to organize, exercise, memorize, select in task-specific way, and modify during learning parametric control of the system.

The notion of control variables (CVs) is strongly related to the notion of parametric control. CVs are those parameters that can be altered by the nervous system in a task-specific way. In some tasks, CVs can be changed in relation to SVs but in other tasks they can be changed independently of SVs or be kept constant. Such freedom of manipulation distinguishes CVs from SVs. By changing CVs, the nervous system may elicit and modulate motor actions, thus taking advantage of natural laws without any knowledge of these laws. This point should be emphasized: the force control theory also assumes that the nervous system takes advantage of natural laws. In contrast, parametric control makes it unnecessary not only an internal imitation but even knowledge of these laws.

For comparison, to transfer the oscillations of a pendulum from one space location to another (Fig. 1), one can simply move its suspension point until the new location is reached following the natural action of mechanical laws. No knowledge of this law is necessary to produce this movement. By repeating this action several times one can improve the movement, for example, by bringing the pendulum to a new location without amplifying or even diminishing oscillations. But this skill may relay on general experience and memory on how the pendulum may react to our manipulations, rather than on intrinsic modeling of its behavior.

With the recognition that control of actions implies changes in parameters of natural law one needs to identify the specific parameters that the nervous system modifies to control posture and movement of the body. The next section briefly reviews the data on such parameter.

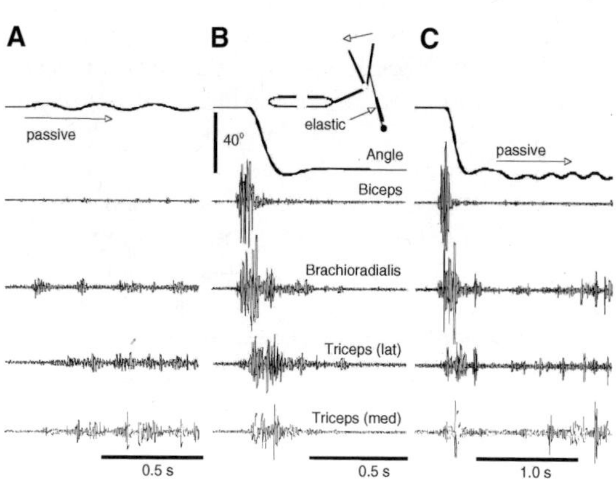

FIGURE 2. Rapid elbow flexion movement (B) and reactions of muscles to passive oscillations at the initial (A) and final (C) positions. Reproduced from [Ostry et Feldman, 03]. Note that the activity of elbow muscles (four lower traces in B) at the initial elbow position is practically zero (background noise level) and, after transient EMG bursts, returns to zero at the final position. Muscles are activated in response to passive oscillations of the arm at the initial (A) and final (C) positions. An elastic connector was used to compensate for the small passive torque of non-active flexor muscles at the initial position of about 140°. The compensation was unnecessary for the final position (about 90°) since it is known that at this position the torque of passive elbow muscles is zero (reproduced from Ostry and Feldman 2003).

THRESHOLD CONTROL AS A FORM OF PARAMETRIC CONTROL

A physiologically well established form of parametric control is shifts in muscle activation thresholds. Specifically, it has been shown that central control levels are able to change a component (λ) of the threshold length value, at which the activity of muscle is initiated. By shifting the thresholds of appropriate muscles, the nervous system produces movement or, if movement is blocked, isometric torques (Asatryan et Feldman, 1965). The threshold control phenomenon can be seen from a simple analysis of fast single-joint movements (Fig. 2).

It may be seen that the EMG activity at the initial position in Fig. 2B is zero but muscles actively reacted to passive oscillations of the arm at this position (Fig. 2 A). This means that motoneurons of arm muscles before movement onset are in a just sub-threshold state. The fact that zero activity and reactions to passive oscillations are also observed at the final position (C) implies that the activation thresholds of motoneurons were reset to this position. The position at which muscles reach their activation thresholds is thus not constant. In other words, the threshold position was reset so that zero muscle activity could be restored at another point in the workspace. This phenomenon is referred to as threshold control. The existence of threshold control follows not only from the simple analysis of the elbow flexion in Fig. 2 but also from many experimental studies in animals and humans, starting from work by Matthews (1959) and Asatryan et Feldman, (1965). The feasibility of threshold control has been demonstrated in many computer simulations of single- and double-joint arm movements. A major significance of the notion of threshold control is that it helps to offer solutions of several motor control problems that remain unsolvable in other approaches. These problems and their solution are reviewed below.

SOLVING SOME MOTOR CONTROL PROBLEMS

Posture-Movement Problem. The notion of threshold control underlies a solution to the classical posture-movement problem of how a movement can occur without triggering resistance of posture-stabilizing mechanisms (for details see (Ostry et Feldman, 2003; Feldman and Latash 2005). Von Holst and Mittelstaedt (1950) formulated a reafference principle, which implies that posture-stabilizing mechanisms, including muscle reflexes, are readdressed to a new posture rather than inhibited when an intentional movement is produced. Specific physiological mechanisms and variables underlying the readdressing were unclear until human studies have shown that the readdressing is achieved by shifting the activation thresholds of appropriate muscles (Asatryan et Feldman, 1965). By shifting muscle activation thresholds, the system readdresses posture-stabilizing mechanisms to a new joint position. The previous position becomes a deviation from the newly specified one, and the same posture-stabilizing mechanisms generate forces that tend to move the joint to the new position. Thus, the system not only eliminates resistance to movement from the previous posture but takes advantage of the posture-stabilizing mechanisms to move to the new posture. By offering a solution to the posture-movement problem, the λ model remains unique since other models of motor control have failed to solve this problem (Ostry and Feldman, 2003).

Problem of Co-Activation. Co-activation of opposing muscle groups is often necessary to speed up and stabilize movements (Feldman et Levin, 1995). Co-activation is also a posture-stabilizing mechanism and, as such, control levels must reset co-activation from the initial to a final posture to prevent resistance to movement. Threshold control solves this problem. Consider, for example, a single joint (for multi-joint movement see (St-Onge and Feldman, 2004) in the absence of a net external torque. Control levels may specify a common threshold angle (r) for all the muscles spanning the joint. At this position, the muscles will be silent (Fig. 2). If a joint is moved passively from position r, muscles stretched by the motion will be activated, whereas the opposing (antagonist) muscles will be activated when the joint is moved passively in the opposite direction. By changing threshold values for the two groups in the same directions, the system shifts the r and thus evokes movement to a new position. By shifting the thresholds of the two muscle group in opposite directions, control levels may surround position r with a zone, in which all muscles may be co-active. The absolute changes in the thresholds for these groups may not be identical as long as they do not influence the net (zero) torque at position r. Thus, in the λ model, co-activation (c) command is defined in terms of muscle activation thresholds of opposing muscle groups, not in terms of EMG activity levels as typically assumed in electrophysiological studies. If command r changes in order to produce an active movement to a new position, the co-activation zone will be automatically shifted with it thus eliminating resistance to the deviation from the initial position. Re-addressed to a new arm position, muscle co-activation contributes to the speed of transition to a new position while increasing damping of the system and thus suppressing terminal oscillations (Feldman and Levin, 1995). In some subjects with hemiparesis, the spatial organization of c commands is deficient, leading to limb instability in these subjects (Levin and Dimov, 1997).

Control of Multiple Muscles. In addition to local biomechanical and reflex factors influencing muscle activation, global factors may be used by the nervous system to control all muscles in a coherent and task-specific way. It has been hypothesized that a virtual or referent (R) configuration of the body determined by muscle recruitment thresholds specified by neural control levels is such a factor. Due to the threshold nature of the R configuration, the activity of each muscle depends on the difference between the actual (Q) and the R configuration of the body. The nervous system modifies the R configuration to produce movement. The referent configuration hypothesis implies that the biomechanical, afferent and central interactions between neuromuscular elements tend to minimize the difference between the Q and R (the *principle of minimization of interactions*). One prediction of this hypothesis is that the Q and R configurations may match each other, most likely in movements with reversals in direction, resulting in a minimum in the electromyographic (EMG) activity level of muscles involved. The depth of the minima is constrained by the degree of co-activation of opposing muscle groups. Another prediction is that EMG minima in the activity of multiple muscles may occur not only when the movement is assisted but also when it is opposed by external forces (e.g., gravity). These predictions have been confirmed for several movements—jumping, stepping in place, sit-to-stand, and head movements (e.g., Lestienne et al., 2000; St-Onge and Feldman, 2004). The concept of referent body configuration has been used in simulation of different movements, including human gait (Gunther and Ruder, 2003).

Guiding Multiple Degrees of Freedom without Redundancy Problems. Threshold control might be helpful in solving the redundancy problem—the problem of how neural control levels guide multiple degrees of freedom of the body to reach a motor goal. The basic idea is the following. Let us assume that some spinal and supraspinal neurons projecting to motoneurons of skeletal muscles of the body, including the extremities may integrate proprioceptive signals from muscles, joint and skin to receive afferent signals, say, about the coordinates of the tip of the index finger (the endpoint) that is typically used to point to targets. The role of these signals will be similar to those of afferents (muscle spindles) that are sensitive to changes in muscle length, except that the recruitment and activity of these neurons will depend not on muscle length but from coordinates of the endpoint. Like for motoneurons, control influences on these neurons can be measured by the amount of shifts in the threshold (referent) coordinates of the endpoint. The difference between the actual and the referent coordinates will determine whether or not such neurons are recruited. These referent coordinates may be shifted by control levels in a frame of reference (FR) associated with the environment to produce *a referent trajectory.* The neuromuscular system will tend to minimize the discrepancy between the actual and referent coordinates forcing the arm and other body segments to move until the endpoint reaches a final position at which a minimum of activity of the neurons in the system in general is reached. After this, the system may compare the output with the desired one. In particular, if the final position of the arm endpoint is different from the desired one, control levels may adjust the referent endpoint trajectory until the final endpoint position coincides with the desired one. Again, although the set of possible configurations for each position of the endpoint is redundant, the minimization process initiated by shifts in the referent coordinates of the endpoint will result in a unique pattern. This configuration pattern can, indeed, vary with repetitions, intentional modifications of the referent pattern, task constraints, including release or restriction in motion of some degrees of freedom (DFs), and history-dependent changes in the neuromuscular system (e.g., due to fatigue).

Action-Producing Frames of Reference. Threshold control implies that neural control levels do not issue instructions on how motoneurons should work in terms of EMG patterns or which forces or torques they should generate. Instead, by determining position-dimension thresholds (threshold lengths, angles, referent configurations, referent position of the effectors) these levels merely pre-determine, in a feedforward way, where, in spatial coordinates motoneurons and muscles should work to produce an action. Specifically, these thresholds can be considered as parameters defining the origins of spatial frames of reference (FRs) in which muscles may be silent or recruited depending on the difference between the respective actual and the threshold position. Other parameters defining, for example, how far the current state of the system from the origin (i.e., metrics), as well as parameters defining the orientation of a given FR in another FR might be also controlled by the nervous system. Changes in such parameters (e.g., shifts in the origin of a FR) result in a change in the activity of motoneurons (actions). Because of these actions, the FRs are called *action-producing* or *physical,* unlike formal, mathematical FRs for which shifting the origin modifies the description of behavior of a system but not behavior itself [Feldman et Levin, 95]. The number of FRs can be enormous but there are certain relationships between them so that the whole set of FRs can be analogous to a tree with hierarchically ordered *major* or *stem* FRs and *branch* FRs embedded in the former. Physical FRs are pre-existing structures so that control levels may chose a FR that is most appropriate for the motor task (*leading* FR) and comparatively rapidly switch to another FR when the task requirement changes, as has been demonstrated for pointing movements to motionless and moving targets (Ghafouri et al., 2002). Indeed, some novel motor tasks may require integration of sensory stimuli not found in the available FRs so that new FRs may be formed during learning.

In conclusion, the notion of threshold control seems fundamental in formulating and solving different problems in the neural control of posture and movement.

References

Asatryan D.G., Feldman A.G. (1965) Functional tuning of the nervous system with control of movements or maintenance of a steady posture: I. Mechanographic analysis of the work of the joint on execution of a postural tasks, Biophysics, 10, 925–935.

Feldman A.G., Levin M.F. (1995) The origin and use of positional frames of reference in motor control. Behavioral Brain Sciences, 18, 723–806.

Feldman A.G. (2005) Equilibrium point control. In Encyclopedic reference of neuroscience. In press.

Feldman A.G., Latash M.L. (2005) Testing hypotheses and the advancement of science: Recent attempts to falsify the equilibrium-point hypothesis. Exp. Brain Res. In press.

Ghafouri M., Archambault P.S., Adamovich S.V., Feldman A.G. (2002) Pointing movements may be produced in different frames of reference depending on the task demand, Brain Research, 929, 117–128.

Gribble P.L., Ostry D.J. (2000) Compensation for loads during arm movements using equilibrium-point control. Exp Brain Res. 135: 474–482.

Gunther M., Ruder H. (2003), Synthesis of two-dimensional human walking: a test of the lambda-model, Biological Cybernetics, 89, 89–106.

Latash ML (1993) Control of Human Movement. Human Kinetics: Urbana, IL.

Lestienne F.G., Thullier F., Archambault P., Levin MF., Feldman A.G. (2000) Multi-muscle control of head movements in monkeys: the referent configuration hypothesis, Neuroscience Letters, 283, 65–68.

Levin M.F., Dimov M. (1997) Spatial zones for muscle coactivation and the control of postural stability, Brain Research, 757, 43–59.

Matthews P.B.C. (1959) A study of certain factors influencing the stretch reflex of the decerebrated cat, J Physiology (London), 147, 547–564.

Ostry D.J., Feldman A.G. (2003) A critical evaluation of the force control hypothesis in motor control, Exp. Brain Res. 153, 275–288.

St-Onge N., Feldman A.G. (2004) Referent configuration of the body: a global factor in the control of multiple skeletal muscles, Exp. Brain Res. 155, 291–300.

Von Holst E., Mittelstaedt H. (1950) Daz reafferezprincip. Wechselwirkungen zwischen Zentralnerven-system und Peripherie, Naturwiss., 37, 467–476, English Transl., The reafference principle, in: The behavioral physiology of animals and man. The collected papers of Erich von Holst. Martin R (translator) University of Miami Press, Coral Gables, Florida, 1971, pp. 139–173.

2. PLANS FOR GRASPING OBJECTS*

David A. Rosenbaum & Rajal G. Cohen
Department of Psychology, Pennsylvania State University, University Park, PA 16802
Ruud G. J. Meulenbroek
Nijmegen Institute for Cognition and Information, University of Nijmegen, Nijmegen, The Netherlands
Jonathan Vaughan
Department of Psychology, Hamilton College, Clinton, NY 13323

Abstract

Through the lens of prehension research, we consider how motor planning is influenced by people's perception of, and their intentions for how to act in, the environment. We review some noteworthy prehension phenomena, including a number of studies from our own labs which demonstrate the "end-state comfort effect," the discovery of sequential effects in motor planning, and the finding that postural end states are known before movements begin. The existence of these phenomena highlights the role that mental representation plays in motor control. We review a recent model of motor control which can account for both perception-related and intention-related features of motor planning.

Introduction

Humanoid robots have made great strides in the last decade. Some modern versions can walk (Sony QRIO, Honda Asimo), vocalize (KRT-v.3—Kagawa University), smile and frown (WE-4R—Waseda University), play the trumpet (Toyota's Partner robot), and hit baseballs at 300 km/hour (University of Tokyo)[1]. However, robots still cannot pass a Turing test for action. A two-year old human can effortlessly pick up objects and inspect them, but robots need extensive intervention to complete such a task. They have difficulty analyzing unfamiliar scenes and deciding how to grasp and manipulate objects of interest. The problem is not merely that robots are unable to achieve basic visual processing or basic motor control. Instead, the problem is that they are poor at planning actions. In order to follow a simple instruction to pick up a rock, a robot must somehow answer questions such as: "From what directions should I approach it? Should I grab it on that outcropping closest to my left? Which posture will allow me to reach it?" and so on. Progress in robotics, as measured by the ability to endow robots with the capacity for autonomous planning, will have achieved a milestone when such questions can be answered without human intervention and when the solutions that are arrived at are indistinguishable from the solutions that normal humans arrive at. When robots can pass such a Turing test for action, we will be able to say with confidence that we truly understand how actions are controlled.

While the foregoing is concerned with robots, this chapter is not about robotics per se. The focus instead is on human action planning, and in particular on how humans plan the grasping of objects. We focus on this aspect of planning because it has been the center of much of our work in the past several years, owing to our belief that the study of plans for grasping

* Chapter prepared for Latash & Lestienne (Ed.), *Progress in Motor Control.* Springer-Verlag.

[1] See http://informatiksysteme.pt-it.de/mti-2/cd-rom/index.html for the state of technologies in human-computer-interfaces

objects provides a window into the nature of planning generally.

The plan for the chapter is as follows. In the first section we describe studies of how grasps depend on the physical properties of the object being grasped. Here we focus on such variables as the size, distance, and direction of the object to be taken hold of and the way these factors affect prehension. In the next section we focus on the way grasps depend on actors' intentions. We focus here on ways that prehension of the same object changes as a function of what one intends to do with the object. In the third part of the chapter, we describe a computational model inspired by and in turn constrained by the results of the findings reviewed in the first and second sections. In the fourth and final section we offer conclusions and consider challenges for future research.

Some disclaimers are in order. Several topics will not be covered here even though they relate to the general topic at hand. These include the neurophysiology of motor planning for grasping and other tasks (e.g., Jeannerod, 1994), studies of haptics for already picked up objects (e.g., Carello & Turvey, 2004), studies of object manipulation that were mainly designed to investigate perception rather than motor planning per se—for example, studies on the possible dissociation of the "what" and "how" visual systems (e.g., Milner & Goodale, 1995), and studies of auditory depth perception (Clifton, Rochat, Litovsky, & Perris, 1991). We omit these topics because this chapter is intended as a selective review of studies from our laboratory. Our research is psychological rather than physiological. We seek a functional analysis of the *software* involved in the formation and implementation of plans for grasping objects rather than a physical analysis of the corresponding *hardware*. As cognitive psychologists, we are interested in uncovering functional principles that, in principle, can be implemented via different hardware—either neural (as in animals) or electro-mechanical (as in robots).

Grasping Based on Perception

How we grasp an object may be influenced by physical properties (of the environment, the object to be grasped, and our own body) which we perceive, and by the effects (on the environment, the object, or ourselves) which we intend to create with that object. We will first consider the influence of the object's physical properties such as size, direction and distance on grasping. In the next section we will discuss how grasping is affected by intention, or by the object's affordances (Gibson, 1979).

Seminal studies of the kinematics of the upper extremity during reaching and grasping objects of different size, direction, and distance were conducted by Jeannerod (1984; 1994), who focused on the transport of the wrist and the opening and closing of the fingers as normal human adults reached for one object at a time. On the basis of Jeannerod's studies as well as the host of studies that were inspired by his work, a number of empirical relations were demonstrated, which we summarize below, drawing on a review we provided of prehension phenomena in an earlier publication (Rosenbaum, Meulenbroek, Vaughan, & Jansen, 2001). Briefly, the phenomena were as follows:

1. When the hand opens to reach for something, the fingers often move much more than the thumb does.
2. The aperture between the fingers and the thumb generally reaches its widest opening in the second half of the movement time (Jeannerod, 1984).
3. The speeding-up phase of a reaching movement is shorter than the slowing-down phase (Jeannerod, 1984).
4. Elbows and shoulders generally display bell-shaped angular velocity profiles (Jeannerod, 1984).
5. Although maximum aperture increases linearly with object size, the slope of the line relating the two is less than 1.0 (Marteniuk, Leavitt, Mackenzie and Athenes, 1990).
6. Maximum aperture occurs relatively later in the reach for a larger object (Marteniuk et al, 1990).
7. Maximum aperture does not depend on the distance to the object being grasped.
8. Maximum aperture tends to increase as movement speed increases (Wing, Turton, & Fraser, 1986).
9. A low-velocity phase is apparent in some reaching movements but not others (Jeannerod, 1981; Wallace & Weeks, 1988). For example, Marteniuk, MacKenzie, Jeannerod, Athenes and Dugas (1987) found that the shape of the velocity profile was different for a reach to a tennis ball than for a reach to a light bulb. The slowing-down phase lasted relatively longer when reaching for a light bulb than when reaching for a tennis ball.

Grasping Based on Intention

The foregoing studies focused on changes in the kinematics of the hand and arm depending on physical properties of objects to be grasped. Object manipulation also depends on what one intends to do with the object or, said another way, with perception of the object's affordances at the moment.

The first investigation that revealed a dependence of prehension on actors' intentions was conducted by Marteniuk, MacKenzie, Jeannerod, Athenes, and Dugas (1987). After demonstrating (as mentioned above) that light bulbs were reached for differently than tennis balls, Marteniuk et al. showed that a single object (in this case a disk) was approached differently depending on whether it was to be thrown or carefully placed after grasping. These findings led Marteniuk et al. to conclude that the kinematics of prehension reflect intentional states.

END-STATE COMFORT

A series of studies in our laboratory was designed to extend this basic observation. The studies were prompted by the sight of a waiter filling glasses with water. The glasses were inverted when they were in their initial, unfilled state. The waiter took hold of each glass with his hand in a thumb-down position. This enabled him to hold the glass with his hand in a thumb-up position when he poured the water into it and also when he placed the filled glass back down on the table. Apparently, the waiter was willing to tolerate initial discomfort when first picking up the glass for the sake of later comfort or control when dealing with it afterward.

To test the generality of this phenomenon and to evaluate possible interpretations given to it, we launched a series of experiments on intentional factors in object manipulation. In the first experiment (Rosenbaum, Marchak, Barnes, Vaughan, Slotta, & Jorgensen, 1990), we asked college students to take hold of a cylinder lying horizontally on a pair of cradles (Figure 1). Two flat target disks lay on either side of the cylinder, one near the left end and one near the right. Participants were asked to reach out with the right hand and grasp the cylinder firmly. There were four conditions: Either the left or the right end of the cylinder was supposed to be placed on the left or right target. The question was what posture participants would adopt upon taking hold of the cylinder.

As shown in Figure 1, the postures that participants adopted depended on what they planned to do with the cylinder. When the *right* end of the cylinder was supposed to be placed down on either target, participants grasped the cylinder with an *overhand* grip, but when the *left* end of the cylinder was supposed to be placed down on either target, participants grasped the cylinder with an *underhand* grip. Thus, the participants anticipated their future bodily states, much as the waiter had done in the restaurant.

Why did subjects modify their grasps as they did? Were they anticipating the comfort of their final

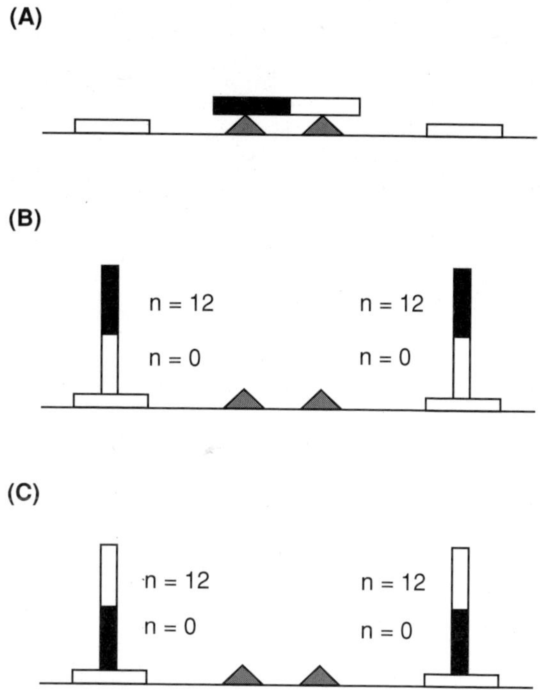

FIGURE 1. (A) Cylinder on the cradle, waiting to be picked up by the participant. (B) Cylinder having been brought to the target with the white side down. The numbers by the black and white ends refer to how many participants grasped at the cradle with the thumb towards that end, when it was to be brought to that target. (C) Cylinder at the target with the black side down. Adapted from Rosenbaum et al., 1990.

postures? To find out, we asked another group of subjects to give ratings of the awkwardness of each of the possible postures for nine tasks similar to those just described (Rosenbaum, Marchak, Barnes, Vaughan, Slotta, & Jorgensen, 1990). The six postures were the overhand and underhand grips of the cylinder in its horizontal orientation, the overhand and underhand grips of the cylinder in its vertical orientation on one target, and the overhand and underhand grips of the cylinder in its vertical orientation on another target. Subjects in the rating study were asked to hold the cylinder in each of these positions and to indicate how awkward the positions felt, using a 5-point scale, where 1 = least awkward up to 5 = most awkward. Rating tasks like this are common in psychology, although to our knowledge they had not been used before in psychological studies of motor control.

The awkwardness ratings that subjects gave are shown in Table 1. In this study subjects had to place the rod in two or three positions in sequence (P1, P2,

TABLE 1. Tasks awkwardness ratings and observed grips (from Rosenbaum et al., 1990).

				Awkwardness Ratings				
Task	*Start*	*Action*	Thumb *Direction*	*P*1	*P*2	*P*3	*Mean*	*Observed*
1	Cradle	White to Red	Black	1.3	1.8	—	1.6	6
			White	3.3	3.1		3.2	0
2	Cradle	Black to Red	Black	1.3	3.1	—	2.2	1
			White	3.2	1.8		2.6	5
3	Red	White to blue	Black	3.1	3.7	—	3.4	0
			White	1.8	1.5		1.7	6
4	Red	Black to Blue	Black	3.1	1.5	—	2.3	5
			White	1.8	3.7		2.8	1
5	Cradle	Black to Red,	Black	1.3	3.1	1.5	2.0	0
		Black to Blue	White	3.3	1.8	3.7	2.9	6
6	Cradle	White to Red,	Black	1.3	1.8	3.7	2.3	5
		White to Blue	White	3.3	3.1	1.5	2.6	1
7	Cradle	Black to Red,	Black	1.3	3.1	3.7	2.7	1
		White to Blue	White	3.3	1.8	1.5	2.2	5
8	Cradle	White to Red,	Black	1.3	1.8	1.5	1.5	6
		Black to Blue	White	3.3	3.1	3.7	3.4	0
9	Red	White to Red	Black	3.1	1.8	—	2.5	2
			White	1.8	3.1		2.5	4

Note. All task descriptions assume starting positions with the black end of the bar in the left end of the cradle or in the red (bottom) disk. P1, P2, P3 denote positions 1, 2, 3, respectively.

and P3). Judged awkwardness at the second of these positions (P2) better predicted grasps at the initial (horizontal) position than at P1 or P3, and better predicted grasps than did overall mean judged comfort; 85 percent of all grasps were in the direction predicted by P2 awkwardness ratings. These results demonstrate that the subjects' choice of grips was not determined by the comfort of their final postures, but the comfort of the second posture to be adopted. Another rating study showed that ratings of *movement* difficulty also failed to predict subjects' grasps as well as end-position comfort ratings. These outcomes led Rosenbaum et al. (1990) and Rosenbaum and Jorgensen (1992) to infer that subjects cared more about final position than end position in motor planning. Accordingly, Rosenbaum et al. (1990) referred to the preference for final comfort over initial comfort as the *end-state comfort* effect.

Several additional studies were performed to evaluate and further elucidate the end-state comfort effect. In these studies (Rosenbaum, Vaughan, Jorgensen, Barnes, & Stewart, 1993) the cylinder that was lifted from the cradle and set down on a target was replaced with a cylinder that was turned from an initial orientation to a final orientation (Figure 2). The cylinder was designed in such a way that the hand could take hold of the cylinder at its axis of rotation. A pointer on one end of the cylinder indicated the cylinder's orientation, and target numbers around the perimeter identified possible orientations to which the cylinder could be brought in each trial (see Figure 2B). Each trial began as in the earlier experiments, with the subject keeping his or her hands by his or her sides. The experiments announced a target to which the pointer should be turned and the subject then reached out with the right hand and grasped the cylinder firmly, rotating it until the pointer was aligned with the target. All required rotations covered 180 degrees.

Figure 3 shows how subjects took hold of the cylinder depending on the orientation to which it would be brought. Subjects were least likely to take hold of the cylinder when the pointer had to be brought to position 4 from position 8 (see Figure 2). As the reader can determine for him or herself, taking hold of the cylinder with the right thumb pointing toward position 8 leaves the arm, after a 180 degree rotation, in a very awkward position. By contrast, taking hold of the cylinder with the right thumb pointing toward position 4 is awkward at first, but the arm ends in a comfortable posture if the cylinder is next rotated 180 degrees. Finding that subjects modify the likelihood of taking hold of the cylinder with the thumb toward the pointer depending on its subsequent position indicates that the end-state comfort effect is a

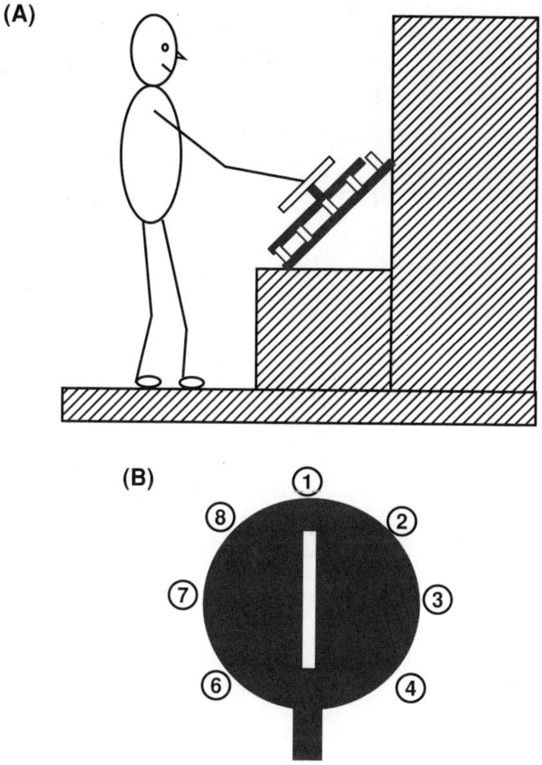

FIGURE 2. Experimental setup with wheel at 45 degrees. From Rosenbaum et al., 1993.

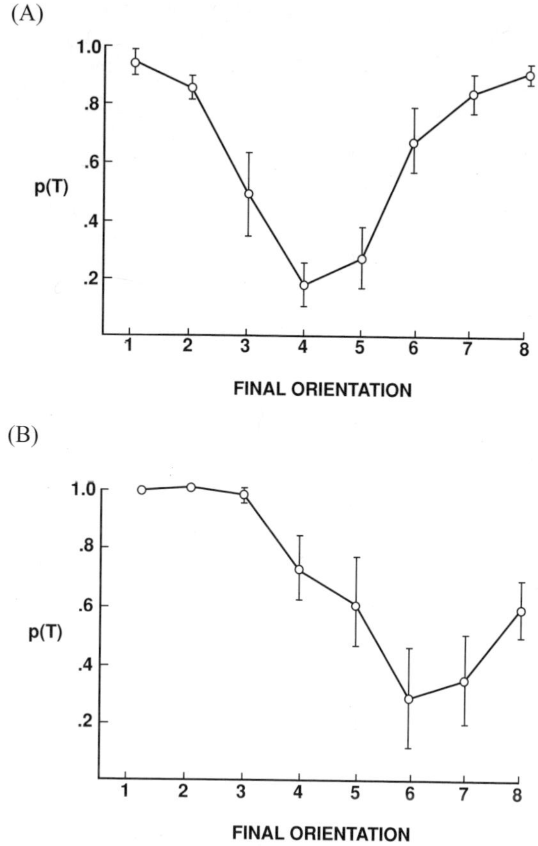

FIGURE 3. Probability p(T) of grasping the cylinder with the thumb toward the pointer-end of the cylinder depending on the required final orientation of the pointer. All the required rotations covered 180 degrees. (A) Data for right-hand turns. (B) Data for left-hand turns. From Rosenbaum et al. (1996).

general phenomenon. Further evidence of the generality of the phenomenon is that it holds for the left arm as well as the right.

GRAVITY

Using the rotating wheel allowed us to test alternative accounts of the end-state comfort effect. One account pertained to the exploitation of potential energy. Perhaps when subjects took hold of the cylinder in initially awkward positions, they knew that they would raise their elbows and that their elbows would drop during the subsequent rotation of the apparatus. Conceivably, subjects exploited gravity to simplify the cost of controlling their arm movements.

To test this hypothesis, Rosenbaum, van Heugten, and Caldwell (1996) took the wheel, which was on a 45 degree tilt in the experiments described above, and placed it on the floor. Subjects sat looking down at the wheel, their feet spread apart and their arms dangling by their sides. In all other respects, the procedure was the same as in the original wheel-turning studies. The data from the wheel-on-the-floor study were virtually the same as the data from the titled-wheel study. As before, subjects freely adopted awkward initial positions to ensure comfortable final positions. Their behavior went against the hypothesis that the end-state comfort reflected a tendency to exploit gravity.

PRECISION

Having the wheel on the floor enabled us to test another possible account of the end-state comfort effect. According to this account, ending in a comfortable posture allows for greater precision than does ending in an uncomfortable posture. To test this hypothesis, Rosenbaum, van Heugten, and Caldwell (1996)

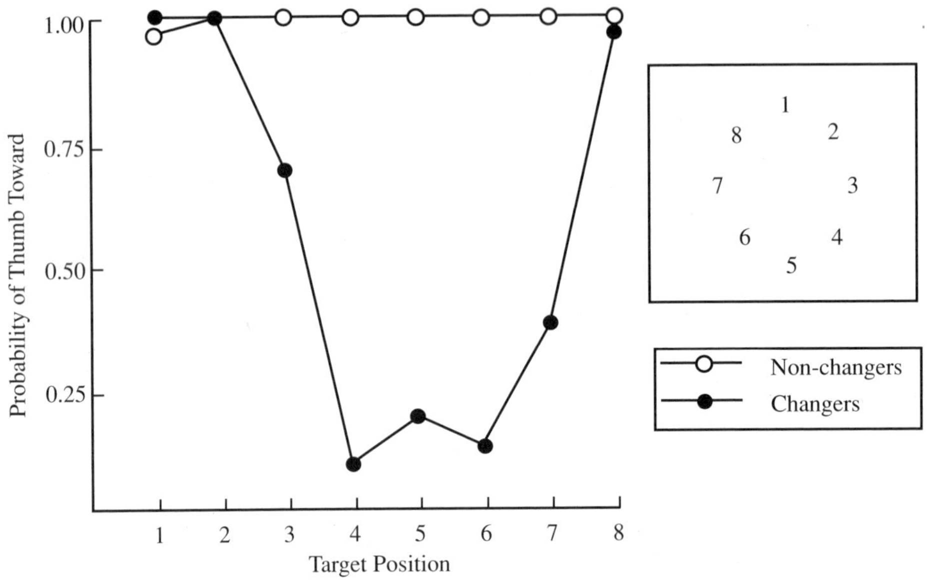

FIGURE 4. Probability of grabbing the cylinder with the thumb toward the pointer-end of the cylinder depending on the required final orientation of the pointer. All the required rotations covered 180 degrees. Inset identifies position numbers. Data for right-hand turns. (Data for left-hand turns looks very similar). From Rosenbaum et al. (1996).

redesigned the wheel on the floor so a bolt dropped into a hole when the wheel reached a target position. This redesign of the apparatus eliminated the need for precise positioning of the wheel near the target locations. The precision hypothesis predicted that the end-state comfort effect would be eliminated in this condition.

The results (Figure 4) were consistent with the precision hypothesis. Whereas virtually all subjects in the previous experiments, where end precision was required, showed the end-state comfort effect, a full half of the subjects in the "dropping-bolt" study did not show the end-state comfort effect. These subjects (the "non-changers") always took hold of the handle with the thumb toward the pointer, which meant that the arm ended up in awkward positions for some required rotations (all of which were 180 degrees, as in the earlier experiments). This *thumb-toward bias* is an interesting example of the use of heuristics in motor planning. The other half of the subjects (the "changers") did show the end-state comfort effect, perhaps because they saw the need for more precise control over the handle's terminal position than was in fact required.

Why would comfortable postures facilitate precision? One possibility is that feelings of discomfort associated with end positions may distract one from attending as fully as needed to precision. A second possibility is that proprioceptive sensitivity is greater at the middle of range of motion than at extreme positions (Rossetti, Meckler, & Prablanc, 1994). A third possibility is that higher torques can be generated at or near the middle of the middle of range of motion than at or near the ends of the range (Winters & Kleweno, 1993). Fourth and finally, people can oscillate the forearm at higher frequencies at or near the middle of the range of motion than at or near the ends, and positions where oscillations are quick may afford more rapid error correction than positions where oscillations are slow (Rosenbaum, van Heugten, & Caldwell, 1996). None of these possibilities is inconsistent with any of the others.

ELASTICITY

Although the precision hypothesis provides the best account of the end-state comfort effect, it is worth mentioning another hypothesis that we considered, partly because the setup used to test it led to the analysis of sequential effects in prehension planning, which is the topic of the next section. According to this other hypothesis, the end-state comfort effect reflected a tendency to store and release elastic energy. The idea was that people effectively wind up the arm and then release it, much as one winds up the rubber band of a toy wooden airplane. Storage and release

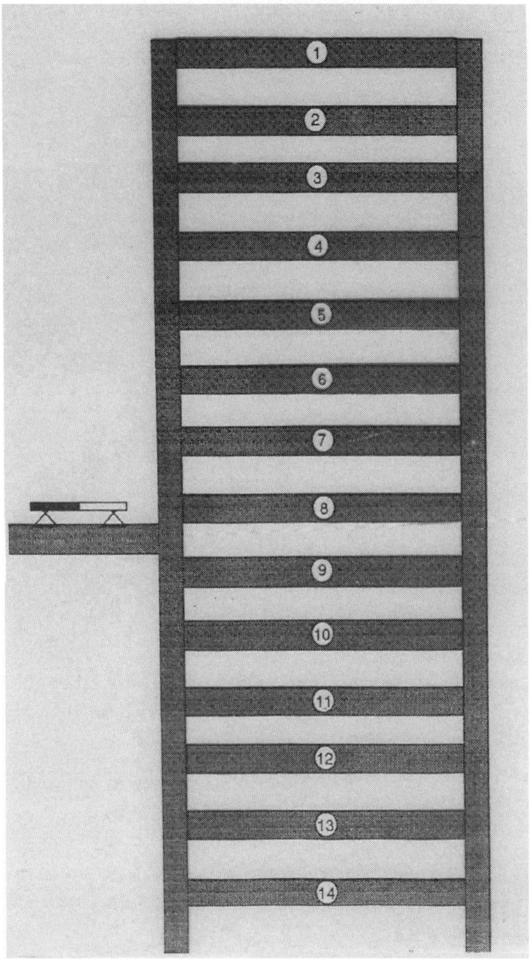

FIGURE 5. Shelf setup used by Rosenbaum and Jorgensen (1992).

of elastic energy is known to play a role in walking and jumping (Alexander & Bennet-Clark, 1977; McMahon, 1984), making conceivable that it plays some role in reaching and grasping.

To test this *elastic energy* hypothesis, Rosenbaum and Jorgensen (1992) devised a task in which the end-state comfort effect would be unlikely to occur if storage and release of elastic energy were actually the source of the effect. In this task (Figure 5) subjects took hold of a cylinder that rested on a cradle and placed the cylinder's left end or right end against a target sitting on the front edge of a shelf. Instructions in each trial indicated which end of the cylinder was to be brought to which target. The main independent variable, aside from which end of the cylinder was supposed to be brought to the target, was the target's height. For most shelves, and especially those that were very high or very low, it was unlikely that the arm could be brought to the necessary position merely by "letting the arm unwind." Accordingly, if the source of the end-state comfort effect was storage and release of elastic energy, the end-state comfort effect would be expected not to occur for these shelves.

Figure 6 shows the results of Rosenbaum and Jorgensen's (1992) "shelf" study. Contrary to the elastic energy hypothesis, the end-state comfort effect was fully replicated at all shelf heights. When subjects, all of whom used the right hand, reached out to take hold of the cylinder to place its *right* end against a target, they were less and less likely to grab hold of the cylinder with an overhand grip the lower the target height. Similarly, when subjects reached out to take hold of the cylinder to place its *left* end against a target, they were less and less likely to grab hold of the cylinder with an underhand grip the lower the target height. This outcome makes sense from the point of view of reducing end-state awkwardness. To push a dowel (even lightly) against a *low* target directly in front of one's body is awkward if the arm is supinated (i.e., if the thumb is away from the target), but to push the same dowel against a *high* vertical target is awkward if the arm is pronated (i.e., if the thumb is toward the target). Both of these positions require the arm to rotate to an extreme degree. The fact that participants in this study exhibited the end-state comfort effect shows that they were aware of this fact. It also argues against the elastic energy hypothesis insofar as the postural transitions that were required in this task were so complex it is unlikely that the simple store and release of elastic energy could underlie the movements. A more plausible interpretation is that subjects sought to adopt end postures that afforded the most efficient means of positioning the cylinder precisely at its targets, as assumed in the precision hypothesis described above.

SEQUENTIAL EFFECTS

The "shelf experiment" of Rosenbaum and Jorgensen (1992) was designed to explore another aspect of the end-state comfort effect besides its possible reliance on elastic energy. In the experiment, target heights were tested in two possible orders—either strictly ascending or strictly descending. Each subject in the experiment was tested in both orders, with half the subjects starting with the ascending order and the other half starting with the descending order. The reason for using ascending and descending orders was to test for *hysteresis*, the tendency for a system to switch from

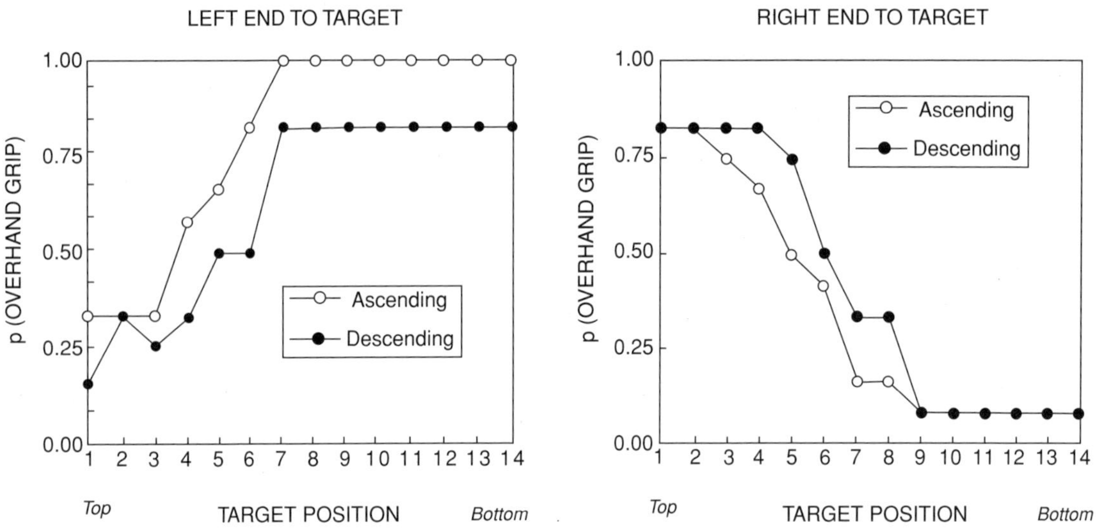

FIGURE 6. Probability of grasping the bar with an overhand grip, depending on the height of the target shelf. From Rosenbaum & Jorgensen, 1992.

one state to another at different values depending on its history.

The data in Figure 6 provide evidence for hysteresis. The height at which subjects switched from an overhand grip to an underhand grip when target heights decreased differed from the height at which subjects switched from an underhand grip to an overhand grip when target heights increased. Thus, there was a sequential effect in subjects' grip choices such that subjects preferred to use the grasp they used before. For this to be true, there had to be a range of heights in which either grasp was tolerable. Rosenbaum and Jorgensen (1992) called this the *range of indifference* for overhand-underhand grasp selection.

MORE EVIDENCE FOR SEQUENTIAL EFFECTS: THE GRASP HEIGHT EFFECT

Do end-state comfort and sequential effects generalize to other grasp tasks? An indication that they do comes from recent work which shifted the focus from choice of overhand or underhand grasps to choice of grasp heights.

An observation in the everyday environment set the stage for this work, much as the observation of the waiter in the restaurant set the stage for the earlier work. One day, the first author walked into his bathroom and saw a toilet plunger standing on the closed toilet lid. He moved the plunger up and to the side to rest it on the counter. After setting the plunger down, he realized he had made an interesting, though unconscious, choice. He had decided where to take hold of the plunger along its length and in so doing had probably anticipated the end state of the plunger, choosing a grasp height that reflected that anticipation. Further informal observations suggested that the measurement of grasp heights could provide a new, potentially sensitive window into plans for grasping objects.

Figure 7 shows the laboratory setup used for the experiments following these initial, informal observations (Cohen & Rosenbaum, 2004). The subject stood before an empty book shelf from which protruded a platform at stomach level. Standing on this "home" platform was a fresh plunger. To the right of the home platform was another protruding "target" platform. The subject was asked to stand with his or her hands by his or her sides and, when ready, to take hold of the plunger with the right hand and move it to the target platform. After doing this, the subject was asked to return his or her hand to his or her side. The performance was videotaped. The height of the home platform was fixed at the middle of the bookshelf. The independent variable was the height of the target platform. The dependent variable was the height along the length of the plunger where the subject took hold of the plunger—what we called the *grasp height*.

Figure 8a shows the result based on freeze-frame analysis of the videotape: The higher the target

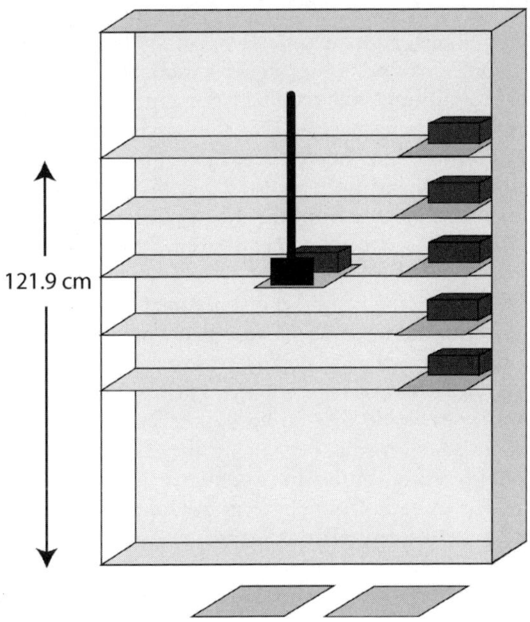

FIGURE 7. Experimental Setup. From Cohen & Rosenbaum, 2004.

platform, the lower the grasp height. The interpretation of this grasp-height effect was not hard to see: Modulating the initial grasp heights so they were inversely related to target heights allowed the hand to come close to the middle of the arm's range of motion at the end of the transport phase of the movement. We concluded that the end-state comfort effect applies in this sort of transport task.

There were sequential effects in this study which, frankly, came as a surprise. In the study, subjects did not just complete a single object transport for each target platform. Instead, after moving the plunger from the home platform to the target platform, they lowered their hands. Next, they reached out again to take hold of the plunger and return it to the home platform, whereupon they lowered their hands once more. Then they repeated the cycle of movements, moving the plunger from the home platform to the target, lowering their hands to their sides, bringing the plunger back to the home platform, and finally resting their hands at their sides. After the second return to home, the experimenter pushed the target platform back into the bookshelf and pulled out the next target platform to be tested. Each of the five target heights was tested in this manner, with the order counterbalanced across subjects. The home platform remained the same throughout the experiment. If grasp heights for the return movements were based entirely on end-state

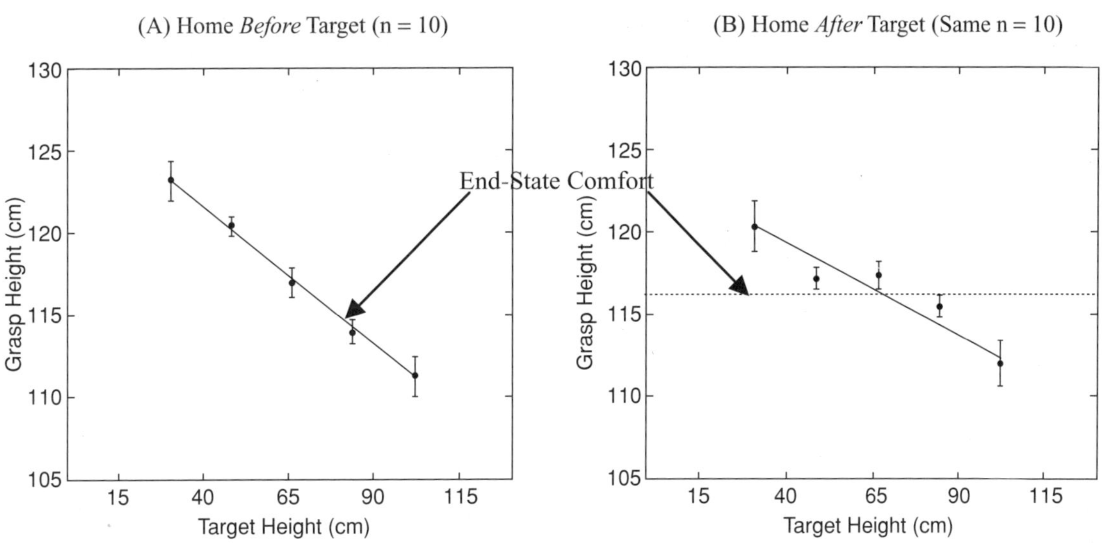

FIGURE 8. Grasp height as a function of target height. (A) Moves from a static home shelf to targets of different heights. (B) Moves from the different targets back to home. From Cohen & Rosenbaum, 2004.

comfort, subjects would ensure that the grasp height back at the home position was fixed, regardless of the height from which the plunger was carried. In fact, as shown in Figure 8b, this was not what happened. Rather than grasping the plunger at a new position that ensured a maximally comfortable end state back at the home platform, subjects grasped the plunger close to where they had grasped it for the home-to-target trip. Thus, subjects exhibited a sequential effect. Further experiments by Cohen and Rosenbaum (2004) confirmed that subjects tried to achieve end-state comfort in first plunger transfers but that their subsequent grasp heights were largely determined by what they had just done. Their bias to grasp the plunger as they had before is similar to what the subjects did in the shelf-height studies of Rosenbaum and Jorgensen (1992). Those subjects also persisted in using overhand or underhand grasps. Insofar as choices of grasp height and choices of overhand-underhand positions both reflect choices of body postures, the results of the two studies indicate that people tend to use the same postures in successive tasks if they can. The discovery of this kind of strategy argues against the idea that movement is optimized from a purely physical perspective (as in theories of minimization of work, torque, jerk, etc.). Instead, the outcome suggests that computational efficiency also matters in movement planning. If the current motor plan is generally satisfactory, continuing to use it is less computationally burdensome than generating a new plan. Expressing this in terms of an American idiom, "If the plan ain't broke, don't fix it!"

TIME TO PLAN GRASPS

What are the real-time processes by which grasps are planned? A reaction-time study by Rosenbaum, Vaughan, Barnes, and Jorgensen (1992) suggested that grasp end states are planned even before reaches are physically initiated. In this study (Figure 9), subjects stood facing a wall-mounted panel with a removable handle with magnetic "feet" protruding from the handle's two ends. The feet rested on two iron disks mounted on the panel. The orientation of the handle depended on which pair of iron disks the handle sat on at the start of each trial. When the subject was ready, as indicated by the fact that s/he pressed his or her hand against a button down by his or her side, a target light appeared beside another pair of iron disks located in one of eight radial positions around the home area. The subject's task was to reach out and pull the handle from its home disks and place it as quickly as possible on the pair of disks designated by the target light. The main dependent measures were the delay between illumination of the target light and release of the start button, the orientation of the hand when it grasped the handle (thumb toward the pointer or away from the pointer), and the time to move the handle from its home position to the target position. Subjects were told to minimize the time between appearance of the target light and placement of the handle on the target position, but they were not told that the time to release the hand from the start button (the reaction time) was separate from the time to carry the handle from the home to the target position (the movement time).

One question behind this experiment was whether subjects would behave in accordance with the end-state comfort effect when they performed under speed pressure. The other question was how subjects' reaction times would depend both on what stimuli they saw and also on what movements they chose to make.

With respect to the first question, as shown in Figure 10, subjects did behave in accordance with the end-state comfort effect. The way they took hold of the handle at its home position anticipated the comfort of their final postures at the targets. Thus, performing under speeded conditions did not eliminate the end-state comfort effect.

With respect to the second question, reaction times for the same home-target disk combinations differed depending on whether subjects grabbed the handle with the thumb toward the pointer or away from the pointer at the home position. That is, even though the choice of hand posture was up to the subjects and even though reaction times did not, in principle, have to change depending on what the chosen hand posture would be, it turned out that reaches culminating in thumb-toward grasps had different reaction times than reaches culminating in thumb-away grasps even when the handle's start position and target position were the same. This outcome suggests that subjects decided even before starting their physical reaches how they would grasp the handle. Furthermore, they made the decision in very little time, judging from the fact that the longer of the two reaction times was only about a third of a second. The discovery that subjects knew how they would end their movements before physically initiating the movements helped set the stage for the model of motor planning that we developed, which is the subject of the next section.

A Model of Motor Planning

The model to be presented next was inspired by and also constrained by the results reviewed above. In what follows, we outline the main claims of the model. Then we indicate how the model accounts for the findings covered earlier. Technical details concerning the model

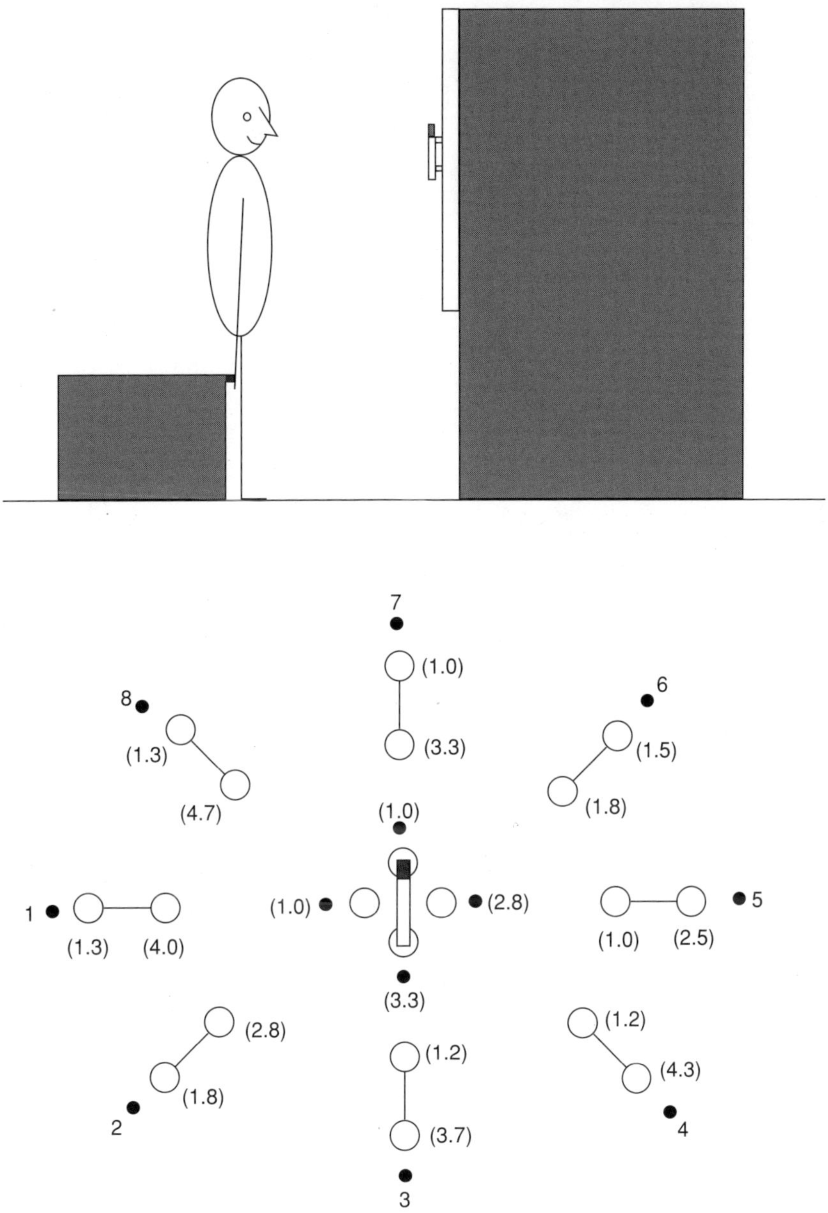

FIGURE 9. Apparatus used in the reaction-time experiment. (Top panel: Side view of a subject, with hand against the start button. The response panel is represented by the white rectangle, and the handle, with the pointer toward the north home position, is represented by the narrow rectangle with the black end on top. Bottom panel: Subject's view of the response panel. The four disks in the center are the four home positions. The handle points to the north home position. The eight pairs of disks surrounding the center are the eight target positions. The small black circles beside each home location and target location represent a light-emitting diode (LED). Target numbers appear beside the target LEDs. In parentheses beside each home and target location is the mean awkwardness rating obtained from raters who held the handle with the thumb toward that location). From Rosenbaum, Vaughan, Barnes, & Jorgensen (1992).

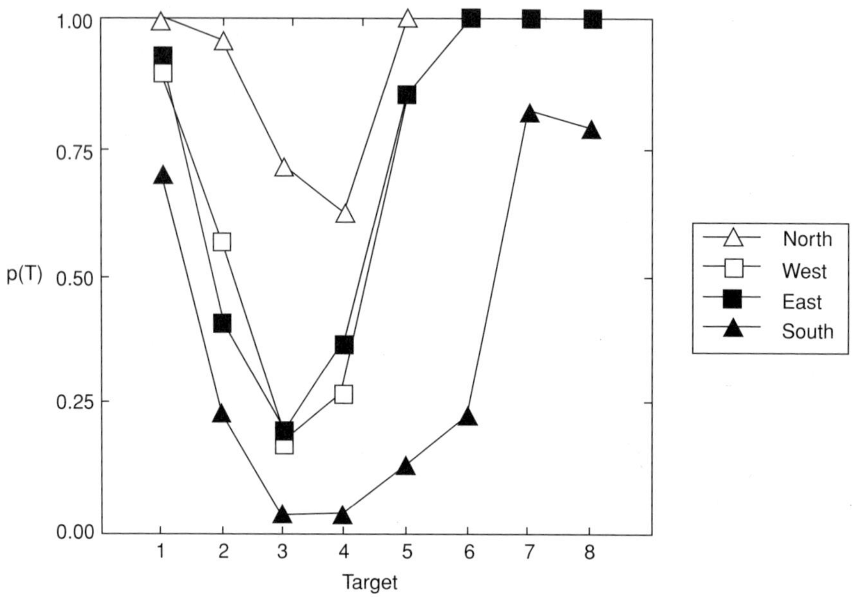

FIGURE 10. Probability, p(T), of grasping the bar with the thumb toward the pointer. From Rosenbaum, D. A., Vaughan, J., Barnes, H. J., & Jorgensen, M. J. (1992).

are suppressed here for the sake of brevity but can be found in Rosenbaum, Meulenbroek, Vaughan, & Jansen (2001). The model is meant to provide a general account of motor planning, not just an account of the planning of grasps. However, we focus below on the model's account of grasping given the focus of this chapter. The model only concerns kinematics, although in principle it could be extended to kinetics. The main claims of the model, along with some supporting evidence for them, are as follows.

1. Movements are planned by first specifying goal postures and then planning trajectories from the start postures to the goal postures. The notion that goal postures are planned before movements are planned fits with the observation that participants in the study of Rosenbaum, Vaughan, Barnes, and Jorgensen (1992) appeared to know what grasps they would end up with even before starting to move. In addition, this claim accords with other data indicating that initial hand speed anticipates the distance to be covered (Atkeson & Hollerbach, 1985; Gordon, Ghilardi & Ghez, 1992). Neither of these findings requires one to conclude that goal postures are planned before movements are planned; they are merely consistent with this idea. However, they do indicate that goal states are known before movements begin, at least for the kinds of movements under consideration. Additional evidence for the hypothesis that goal postures are normally planned before movements comes from neurophysiological evidence that prolonged microstimulation of specific areas in the primary and premotor cortex of monkeys leads to adoption of characteristic postures regardless of the monkey's initial posture (Graziano, Taylor & Moore, 2002). The discovery of such "posture neurons" is consistent with the view that there is a way to specify body positions prior to the initiation of motion, a concept that originates with the equilibrium-point hypothesis of motor control (Asatryan & Feldman, 1965).

2. Goal postures and movement trajectories are chosen with respect to a constraint hierarchy—a prioritized list of constraints whose rank order (most important constraint down to least important constraint) defines the task to be performed. A typical constraint is generating movements that entail acceptable levels of effort, where the acceptable levels depend on the task (e.g., weight lifting can entail more effort than feather dusting). Another typical constraint is generating movements that ensure adequate clearance around obstacles. The amount of clearance also depends on the task. Large clearances are needed if

dangerous objects must be avoided, whereas low or no clearances can be used when objects should be touched.

3. Movements are assumed to have bell-shaped tangential velocity profiles and to be straight lines through joint space from the starting posture to the goal posture unless different trajectories are needed. The assumption that movements have bell-shaped tangential velocity profiles has been supported in many studies (Hogan, 1984; Morasso, 1981). The assumption that movements are, by default, straight-line paths through joint space is motivated by the idea that goal postures are specified before movements, so movements are viewed in the theory as being, in effect, interpolations from start to goal postures. Straight-line motions through joint space have been observed (Soechting & Lacquaniti, 1981), although straight-line movements through extrinsic space have been observed more often (Abend, Bizzi & Morasso, 1982). Because movement trajectories can be shaped in the theory (see item 5 below), it is possible to deliberately generate straight-line movements through extrinsic space using the theory's computations.

4. Movements are evaluated via forward kinematics before being performed to determine if their default forms need to be changed. Reliance on feedforward modeling is well established for movement control (see e.g. Wolpert & Flanagan, 2001). A default movement may be judged unacceptable if it would result in a collision or if the shape differs from a desired shape, as in writing or dancing.

5. If a default movement is rejected, it is combined with another movement to make an acceptable *compound* movement. The movement with which the main movement is combined is assumed to be a back-and-forth movement that goes from the starting posture to a "bounce posture" and back to the starting posture. The bounce posture is selected by using a constraint hierarchy, just as the goal posture is (Vaughan, Rosenbaum, & Meulenbroek, 2001). The direction and distance of the bounce posture from the starting posture affects the curvature of the compound movement. The main movement and the back-and-forth movement are assumed to start and end together. Combining movements is a well established capability in the study of motor control (Pigeon, Yahia, Mitnitski, & Feldman, 2000).

6. Goal postures are assumed to be selected through a two-stage process. The first stage consists of determining which stored posture—the last m adopted goal postures are assumed to be stored—is most promising for the task at hand, as defined with respect to the constraint hierarchy. The second stage consists of "tweaking" that most promising stored posture via a diffusion process (i.e., generating candidate goal postures similar to the most promising stored posture). This aspect of the theory was supported by Rosenbaum and Jorgenson's (1992) and Cohen and Rosenbaum's (2004) discovery of sequential effects in the postures chosen for transport tasks. Of all the postures that were candidate goal postures, the one that survives the deepest cuts down the constraint hierarchy becomes the goal posture. According to this claim, recently adopted goal postures can be most useful if they are quite similar to goal postures that need to be adopted for the present task. Thus, some of the benefit of "warming up" is explained by appealing to the prevalence of stored goal postures that may be useful for a particular task. Having stored goal postures that satisfy many constraints for the task reduces the duration and/or depth of the diffusion around the most promising stored posture. The theory does not assume that movements per se are learned, because such an assumption would be unnecessary. Consistent with this claim, it is well known that end positions of movements are remembered better than movements themselves (see Smyth, 1984, for review, and Rosenbaum and Dawson, 2004, for recent discussion).

7. Regarding prehension, no special assumptions are required. Hand and arm positions are treated like any other kind of postures. The one exception is that in the simulations of reach and grasp movements reported by Rosenbaum, Meulenbroek, Vaughan, and Jansen (2001) and Meulenbroek, Rosenbaum, Jansen, Vaughan & Vogt, (2001), the hand was treated as a sub-unit of posture space. Partitioning the hand and arm this way was introduced for computational convenience only, although it is interesting that others have likewise entertained the hypothesis that the hand may be represented as a hierarchical sub-unit of the arm. This hypothesis has been advanced both in motor control (Jeannerod, 1984; Klatzky et al, 1987) and in perception (Marr, 1982; see Figure 11).

8. In grasping objects, moving directly (in joint space) from a starting posture to a goal posture that achieves a precision or power grip on the object would almost always result in a collision with the object before the grip is achieved. However, the model does not need a special mechanism for making collision-free movements to grip postures. It simply exploits the obstacle-avoiding mechanism (item 5) by which an unsatisfactory default (direct) movement is combined with another movement to make an effective compound movement (Vaughan, Rosenbaum, & Meulenbroek, 2001) to attain the grip posture without

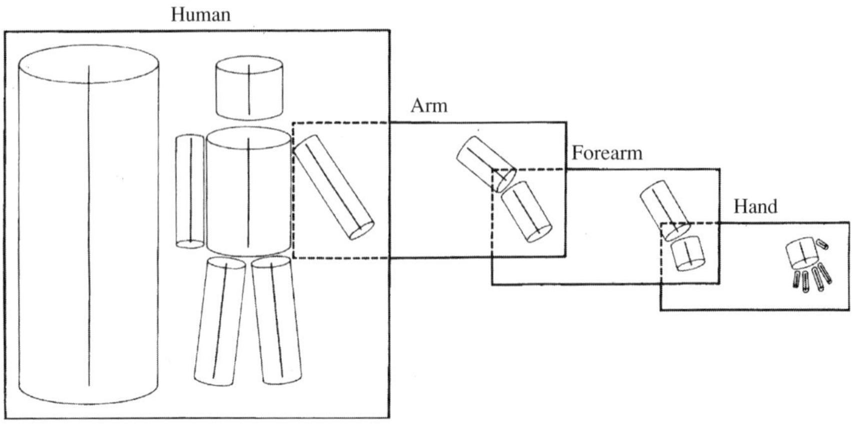

FIGURE 11. Hierarchical composition of the human body thought to be used in perceptual analysis of body forms. From Marr, D. (1982). *Vision.* San Francisco: W. H. Freeman.

colliding. Thus, no additional assumptions are required for the model to accommodate the obstacle-avoiding dimension of reaching to grasp an object (Rosenbaum, Meulenbroek, Vaughan, & Jansen, 2001).

How do these ideas come together in actual simulations of grasping movements? Figure 12 shows just one of the simulations generated on the basis of the model. The figure shows an artificial creature reaching out to take hold of an object. Also included in this figure is a panel showing how the wrist tangential velocity and distance between the thumb and index finger changed together over time. The two panels on the right side of the figure show angular velocity profiles for the joints involved. Altogether, the movement is realistic, both at the level of informal observation of the animation and at the level of more detailed, quantitative examination. Indeed, the features of prehension listed above in the section called **Grasping Based on Perception** are all accounted for with the model. For detailed expositions of the accounts, see Rosenbaum, Meulenbroek, Vaughan, and Jansen (2001) and Meulenbroek, Rosenbaum, Jansen, Vaughan, and Vogt (2001). For extensions of the model to the understanding of grasping in the context of spasticity, see Meulenbroek, Rosenbaum, and Vaughan (2001). The article by Rosenbaum, Meulenbroek, Vaughan, and Jansen (2001) also covers other aspects of motor performance, not specifically tied to grasping, which the model handles. Among these aspects are immediate compensation for changes in joint mobility, changes in the relative contributions of the joints depending on the required speed of movement, and the importance of accurate information about one's starting position.

Conclusions

Through the lens of prehension research, we have considered how motor planning is influenced by perceptions of the environment and by intentions of the actor. We reviewed some noteworthy prehension phenomena, including a number of studies from our own labs. In particular, three lines of research from our labs were especially relevant: (1) the phenomenon we call "end-state comfort"; (2) the discovery of sequential effects in motor planning; and (3) the finding that postural end states are known before movements begin. The existence of these phenomena highlights the important role that mental representation plays in motor control above the most basic level. We outlined a model of motor control that can account for both perception-related and intention-related features of motor planning.

Regarding the theory, we also allow for the possibility that the planning of movements can be bi-directional: choice of movement can reciprocally influence the choice of goal posture (Kawato, 1996). So far, we have applied the theory *quantitatively* to 2-dimensional aspects of prehension and only *qualitatively* to 3-dimensional aspects. We would like to extend the model to account for 3-dimensional moves. We also hope to extend the theory to include kinetics, not just kinematics. Here it is relevant that even babies learn to anticipate the forces required to lift objects based on their experience with the object's weight in

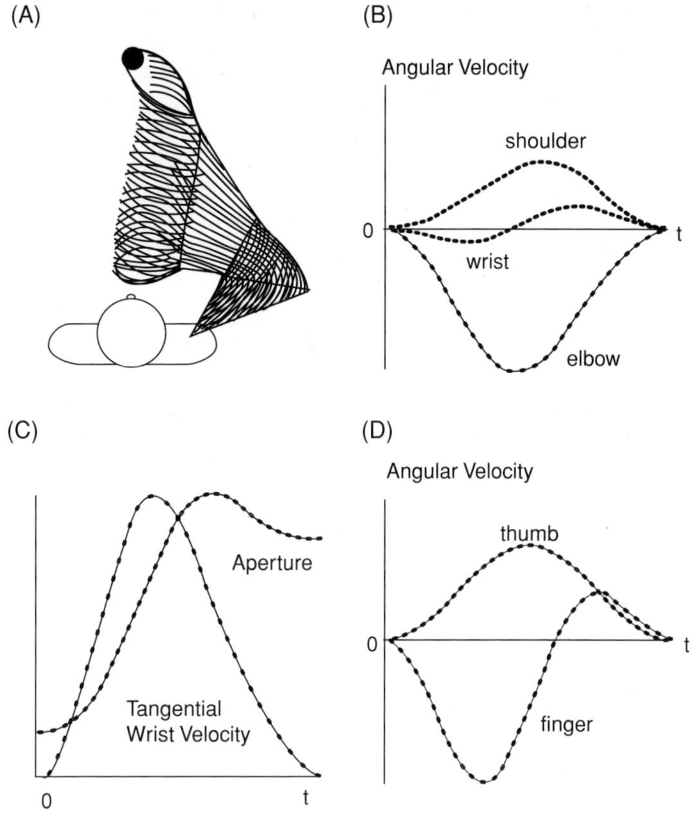

FIGURE 12. Simulated reach and grasp based on the posture-based motion planning model. (A) Stick figure animation. (B) Shoulder, elbow, and wrist angular velocity profiles. (C) Wrist tangential velocity and thumb-index finger aperture profiles. (D) Thumb and index finger angular velocity profiles. From Rosenbaum, Meulenbroek, Vaughan, & Jansen (2001).

repeated lifts. If the weight of the object is suddenly changed, the baby will lift it "too hard" (Gachoud, Mounoud, Hauert, & Viviani, 1983).

Our theory has been criticized for its computational complexity (Smeets & Brenner, 2002), but there is a tradeoff between complexity and number (or range) of phenomena accounted for. We believe that the large number and variety of phenomena successfully accounted for by our theory justify its relative complexity. To our knowledge, no simpler model exists that accounts for the phenomena described here. If others develop such a model, that would be a welcome contribution to progress in motor control.

Author Notes

The work described in this chapter was supported by grant SBR-94-96290 from the National Science Foundation, grants KO2-MH0097701A1 and R15 NS41887-01 from the National Institute of Mental Health, and the Research and Graduate Studies Office of The College of Liberal Arts, Pennsylvania State University. Correspondence should be sent to David A. Rosenbaum (DAR12@PSU. EDU) at the Department of Psychology, Moore Building, Pennsylvania State University, University Park, PA 16802.

References

Abend, W., Bizzi, E., & Morasso, P. (1982). Human arm trajectory formation. *Brain, 105*, 331–348.

Alexander, R. M. & Bennet-Clark, H. C. (1977). Storage of elastic strain energy in muscle and other tissues. *Nature, 265*, 114–117.

Asatryan, D. G., & Feldman, A. G. (1965). Functional tuning of the nervous system with control of movement or maintenance of a steady posture. 1. Mechanographic analysis of the work of the joint on execution of a postural task. *Biophysics, 10*, 925–935.

Atkeson, C. G., & Hollerbach, J. M. (1985). Kinematic features of unrestrained arm movements. *The Journal of Neuroscience, 5*, 2318–2330.

Carello, C. & Turvey, M. T. (2004). Physics and psychology of the muscle sense. *Current Directions In Psychological Science, 13*, 25–28.

Clifton, R., K., Rochat, P., Litovsky, R. Y., & Perris, E. E. (1991). Object representation guides infant reaching in the dark. *Journal of Experimental Psychology: Human Perception and Performance, 17*, 323–329.

Cohen, R. G. & Rosenbaum, D. A. (2004). Where objects are grasped reveals how grasps are planned: Generation and recall of motor plans. *Experimental Brain Research, 157*, 486–495.

Gachoud, J. P., Mounoud, P., Hauert, C. A., & Viviani, P. (1983). Motor strategies in lifting movements: A comparison of adult and child performance. *Journal of Motor Behavior, 15* (3), 202–216.

Gibson, J. J. (1979). *An ecological approach to visual perception.* Boston, Houghton Mifflin.

Gordon, J., Ghilardi, M. F., & Ghez, C. (1992). In reaching, the task is to move the hand to a target. *Behavioral and Brain Sciences, 15*, 337–338.

Graziano, M. S., Taylor, C. S. R., & Moore, T. (2002). Complex movements evoked by microstimulation of precentral cortex. *Neuron, 34*, 841–851.

Hogan, N. (1984). An organizing principle for a class of voluntary movements. *The Journal of Neuroscience, 4*, 2745–2754.

Jeannerod, M. (1984). The timing of natural prehension movement. *Journal of Motor Behavior, 26*, 235–254.

Jeannerod, M. (1994). The representing brain: Neural correlates of motor intention and imagery. *Brain and Behavioral Science, 17*, 187–245.

Kawato, M. (1996). Bidirectional theory approach to integration. In: T. Inui and J.L. McClelland (Eds), *Attention and Performance XVI* (pp 335–367). MIT Press, Cambridge (MA), USA.

Klatzky, R. L., McCLoskey, B., Doherty, S., Pellegrino, J. & Smith, T. (1987). Knowledge about hand shaping and knowledge about objects. *Journal of Motor Behavior, 19*, 187–213.

Marr, D. (1982). *Vision.* San Francisco: W. H. Freeman.

Marteniuk, R. G., MacKenzie, C. L., Jeannerod, M., Athenes, S., & Dugas, C. (1987). Constraints on human arm movement trajectories. *Canadian Journal of Psychology, 4*, 365–378.

Marteniuk, R. G., Leavitt, J. L., MacKenzie, C. L., & Athenes, S. (1990). Functional relationships between grasp and transport components in a prehension task. *Human Movement Science, 9*, 149–176.

McMahon, T. A. (1984). *Muscles, reflexes, and locomotion.* Princeton, NJ: Princeton University Press.

Meulenbroek, R. G. J., Rosenbaum, D. A., Jansen, C.,Vaughan, J., & Vogt, S. (2001). Multijoint grasping movements: Simulated and observed effects of object location, object size, and initial aperture. *Experimental Brain Research, 138*, 219–234.

Meulenbroek, R. G. J., Rosenbaum, D. A., Jansen, C., Vaughan, J., & Vogt, S. (2001). Multijoint grasping movements: Simulated and observed effects of object location, object size, and initial aperture. *Experimental Brain Research, 138*, 219–234.

Meulenbroek, R. G. J., Rosenbaum, D. A., & Vaughan, J. (2001). Planning reaching and grasping movements: Simulating reduced movement capabilities in spastic hemiparesis. *Motor Control* 5, 136–150.

Milner, A. D. & Goodale, M. A. (1995). *The visual brain in action.* New York: Oxford University Press.

Morasso, P. (1981). Spatial control of arm movements. *Experimental Brain Research, 42*, 223–227.

Pigeon, P., Yahia, L. H., Mitnitski, A. B., & Feldman A. G. (2000). Superposition of independent synergies during pointing movements involving the trunk in the absence of vision. *Experimental Brain Research, 131*(3), 336–49.

Rosenbaum, D. A. & Dawson, A. (2004). The motor system computes well but remembers poorly. *Journal of Motor Behavior, 36*, 390–392.

Rosenbaum, D. A., van Heugten, C., & Caldwell, G. C. (1996). From cognition to biomechanics and back: The end-state comfort effect and the middle-is-faster effect. *Acta Psychologica, 94*, 59–85.

Rosenbaum, D. A. & Jorgensen, M. J. (1992). Planning macroscopic aspects of manual control. *Human Movement Science, 11*, 61–69.

Rosenbaum, D. A., Marchak, F., Barnes, H. J., Vaughan, J., Slotta, J., & Jorgensen, M. (1990). Constraints for action selection: Overhand versus underhand grips. In M. Jeannerod (Ed.) *Attention and performance XIII* (pp. 321–342). Hillsdale, NJ: Erlbaum.

Rosenbaum, D. A., Meulenbroek, R. G., Vaughan, J., & Jansen, C. (2001). Posture-based motion planning: Applications to grasping. *Psychological Review, 108*, 709–734.

Rosenbaum, D. A., Vaughan, J., Barnes, H. J., & Jorgensen, M. J. (1992). Time course of movement planning: Selection of hand grips for object manipulation. *Journal of Experimental Psychology: Learning, Memory, and Cognition, 18*, 1058–1073.

Rosenbaum, D. A., Vaughan, J., Jorgensen, M. J., Barnes, H. J., & Stewart, E. (1993). Plans for object manipulation. In D. E. Meyer & S. Kornblum (Eds.), *Attention and performance XIV—A silver jubilee: Synergies in experimental psychology, artificial intelligence and cognitive neuroscience* (pp. 803–820). Cambridge: MIT Press, Bradford Books.

Rossetti, Y., Meckler, C., & Prablanc, C. (1994). Is there an optimal arm posture? Deterioration of finger localization precision and comfort sensation in extreme arm-joint postures. *Experimental Brain Research, 99*, 131–136.

Smeets, J. B. J., & Brenner, E. (2002). Does a complex model help to understand grasping? *Experimental Brain Research, 144*, 132–135.

Smyth, M. M. (1984). Memory for movements. In M. M. Smyth & A. M. Wing (Eds.), *The psychology of human movement* (pp. 83–117). London: Academic Press.

Soechting, J. F., & Lacquaniti, F. (1981). Invariant characteristics of a pointing movement in man. *Journal of Neuroscience, 1*, 710–720.

Vaughan, J., Rosenbaum, D. A., and Meulenbroek, R. G. J. (2001). Planning reaching and grasping movements: The problem of obstacle avoidance. *Motor Control, 5*, 116–135.

Wallace, S. A. & Weeks, D. L. (1988). Temporal constraints in the control of prehensile movement. *Journal of Motor Behavior, 20*, 81–105.

Wing, A., Turton, A., & Fraser, C. (1986). Grasp size and accuracy of approach in reaching. *Journal of Motor Behavior, 18*, 245–260.

Winters, J. M. & Kleweno D. G. (1993). Effect of initial upper-limb alignment on muscle contributions to isometric strength curves. *Journal of Biomechanics, 26*, 143–153.

Wolpert, D. R., & Flanagan, J. R. (2001). Motor prediction. *Current Biology, 11* (18), R729–R732.

3. ADHERENCE AND POSTURAL CONTROL: A BIOMECHANICAL ANALYSIS OF TRANSIENT PUSH EFFORTS

Simon Bouisset

Laboratoire de Physiologie du Mouvement, Université Paris-Sud, 91405 ORSAY, France

Serge Le Bozec

Laboratoire de Physiologie du Mouvement, Université Paris-Sud, 91405 ORSAY, France; U731 INSERM / UPMC

Christian Ribreau

Laboratoire de Biomécanique et Biomatériaux Ostéo-Articulaires, Université Paris 12 – Val de Marne, 94010 CRETEIL, France

Abstract

This chapter focuses on the question of the interface between the body and its physical environment, namely adherence and friction. First, a short survey of literature is presented and some basic statements on adherence reviewed. They help define the adherence constraints associated with different motor tasks. Then, a new paradigm is presented, the transient push paradigm, which offers manifold facilities. In particular, it makes it possible: i) to exert transient external force in the absence of external movement; ii) to divide the body into a focal and a postural chain; and iii) to manipulate the surface contacts between the body and its supports, without perturbing body balance.

The chapter is documented with recent results on transient isometric pushes performed under two conditions of surface contact. A biomechanical model is presented. Based on an experimental recording of the main terms of the model, it is concluded that transient muscular effort induces dynamics of the postural chain. These observations support the view that there is a postural counter-perturbation, which is associated with motor acts. Changing ischio-femoral contact has been proven to modify postural chain mobility, which appears to be a key factor of performance.

The influence of adherence was considered from the adherence ratio, that is, $\mu = R_T/R_N$ (with μ being the adherence ratio, R_T and R_N, the instantaneous tangential and normal reactions at the contact surface). It was found to evolve, during the course of the effort, up to a certain value, which is close to the coefficient of friction to within a security margin, at the seat contact surface, at least. Lastly, the adherence effects on motor programming are highlighted, and the possibility of considering the centre of pressure as the postural control variable is discussed. It is proposed that the instantaneous adherence ratio, with reference to the coefficient of friction, might be one of the rules for controlling muscle activation to accomplish voluntary efforts, when there is the risk of loosing balance.

Keywords: Postural dynamics; ramp push efforts; adherence, motor control.

When they move, human, as well as animals, have to comply with mechanical rules, known as laws of dynamics ("Newton's laws"). The forces taken into consideration are those which are external to the system. For example, when the human body is considered as a whole, the external forces are limited to gravity and the reactions are developed at the interface with the physical environment, primarily the ground reaction if the movement is performed on the earth track. Moreover, it is well known that the ground reaction, and, more generally, the reactions induced by the contact areas, depends on their physical properties, that is rigidity and adherence.

The aim of this chapter is to highlight the interactions between postural dynamics and **adherence**, and to discuss their effects on motor control. It is documented with recent results on the transient push paradigm, which is considered a "pure" kinetic motor act, in that there is no hand movement, even though muscular effort varies at each instant.

1. *Adherence, Friction and Postural Dynamics*

A short survey of literature will be useful before reviewing some basic statements on adherence and the articulated body chain.

1.1. ADHERENCE AND MOTOR ACTS: A LITERATURE SURVEY

The problem of adherence has been considered according to two main biomechanical viewpoints. The first set of research was more practical. It included gross body movements, such as exertion of push/pull force (Gaughran and Dempster, 1956; Whitney, 1957; Kroemer, 1974; Grieve, 1979), walking (Carlsöö, 1962, Lanshammar and Strandberg, 1981; Strandberg, 1983; Tisserand, 1985), running (see Nigg, 1986, for a review) and ice skating (de Koning and Van Ingen Schenau, 2000). Most of the studies on push/pull forces focused on maximal force exertion, that is static conditions, and aimed at defining the most efficient ones. Studies on locomotion considered necessarily dynamic conditions. Those on walking were conducted with the aim of measuring floor/shoe slip resistance in order to prevent slipping, and to prevent fall-related injuries, and those on running and ice skating, with the aim of improving performance.

The second set of research focused on the mechanisms controlling the contact forces at the hand, during both "static" and "dynamic" efforts (Johansson and Westling, 1984; see Wing, 1996, for a review). The manual efforts under consideration included transient grip force paradigms. Prehension forces are limited to the grip force and the load force (that is the object's weight). The results stressed the influence of friction on the motor act, and acknowledged the importance of a safety margin, in order to prevent slipping (Flanagan and Wing, 1995; Westling and Johansson, 1984). More precisely, the grip force would have to be calibrated in relation to the load force. It was concluded that the coefficient of friction might be implemented in the motor program.

The prehension studies focused on the efforts exerted at the hand level, contrary to the gross body movement studies, which considered every reaction forces at the interface between the subjects and the physical environment. In order to study the question, a series of experiments on transient push efforts was recently initiated. Biomechanical (Bouisset *et al.*, 2002; Le Bozec *et al.*, 1996; 1997; Le Bozec and Bouisset, 2004) and EMG (Le Bozec *et al.*, 2001; Le Bozec and Bouisset, 2004) data were considered, and a biomechanical model was elaborated (Bouisset *et al.*, 2002). The main biomechanical results are presented later on, with the aim of stressing their contribution to the motor control approach.

1.2. COEFFICIENT OF FRICTION AND ADHERENCE RATIO

The Coefficient of Friction (CoF) is defined at the slipping limit by the well-known relationship:

$$R^*_T = \mu^* R^*_N \qquad (1)$$

where μ^* is the **coefficient of friction**, R^*_T the tangential reaction (or friction force), and R^*_N the normal reaction at the contact surface. The coefficient of friction varies according to the properties of the interface, and the risk of slipping increases as μ^* decreases.

In order to evaluate adherence, and consequently the risk of slipping, an Adherence Ratio (AR) can be defined, which is:

$$R_T = \mu R_N \qquad (2)$$

where μ is the **adherence ratio**, and R_T and R_N, the instantaneous tangential and normal reactions at the contact surface (Fig. 1).

During locomotion, AR was also called "friction use" by Strandberg (1983), and was defined by the ratio between the tangential and vertical ground reactions. However, during prehension, the inverse of AR was usually considered, that is, the ratio of the grip force (that is, normal force) to the load force (that is, tangential force). It was called the "slip ratio" (Johansson and Westling 1987).

Adherence and friction are close companions, because the coefficient of friction is the boundary mark of the adherence ratio ($\mu \leq \mu^*$). However, AR is not a measure of CoF since, by this very fact, it varies under

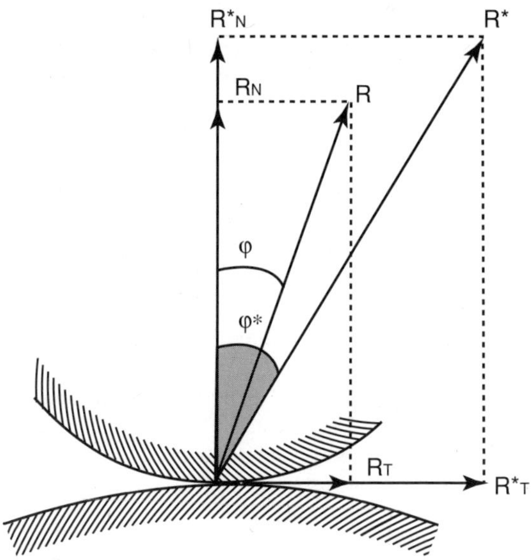

FIGURE 1. Coefficient of friction and adherence ratio. R^*_T, tangential reaction (or friction force), and R^*_N, normal reaction at the contact surface, are the reactions at the slipping limit; φ^* is the friction angle. R_T and R_N are the actual tangential and normal reactions at the contact surface; φ is the actual angle of adherence Adherence ratio (that is μ) reaches the limit of slipping (that is μ^*) when R_T increases, and/or when R_N decreases. There is no slipping as long as as $\varphi < \varphi^*$.

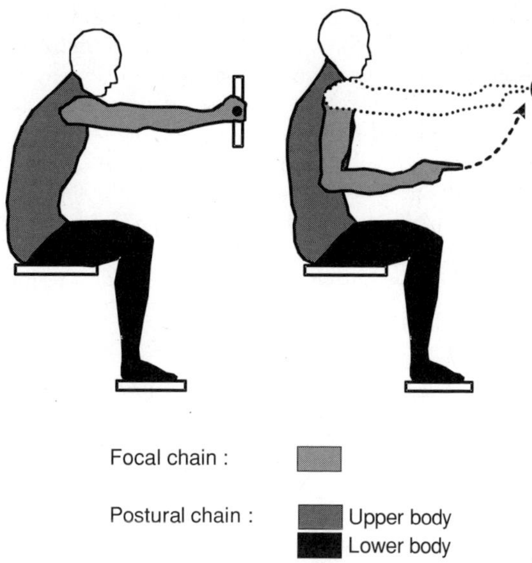

FIGURE 2. Focal and postural chains. The partitioning of the body between a focal and a postural chain is illustrated in pushing (left) and pointing (right) tasks.

CoF until slipping occurs. However, AR reflects how the CNS takes into account the contact forces between the body and its physical environment in order to perform the motor act efficiently. In addition, it can be assumed that the higher the CoF, the higher the AR, that is, the more the contact forces are put into play.

More generally, for a given interface, the CoF value appears to delimit two motor behaviours: it separates the domain where voluntary action can proceed in accordance with the primary intent, from the domain where it is perturbed by unexpected slipping and a possible fall.

1.3. REACTION FORCES AND POSTURAL CHAIN DYNAMICS

From a biomechanical viewpoint, the skeleton's structure allows the modelling of the human body as an articulated chain of rigid solids, which are actuated in relation to each other. The forces (and torques) are transmitted between the segment(s) the subject intentionally mobilizes and the distal one(s) and between these and the physical supports. As a consequence, an intended movement involves a perturbation of body balance, as has been suggested by several neurologists since the turn of the last century (see, for example, André-Thomas, 1940).

This is why it has been proposed that the articulated body chain be divided into two functional parts (Bouisset and Zattara, 1981 and 1983). One, the **focal chain**, would be directly in charge of voluntary movement, that is, of the task movement the subject intends to perform. The other, the **postural chain**, includes the rest of the body. It would be responsible for the stabilizing action, which must be opposed to the balance perturbation provoked by voluntary movement. This counter-perturbation is necessary in order to perform the task efficiently (Bouisset and Zattara, 1981; Friedli *et al.*, 1988).

For example (Fig. 2), when pointing at a target with the upper limb, this limb clearly represents the focal chain. Similarly, when pushing on a bar, the intended push force originates from the shoulder muscles and is transmitted to the bar through the upper limbs, which constitute the focal chain. The chain located between the shoulders and the ground is the postural chain. Again, it is easy to divide the postural chain into two parts, particularly when the effort is performed in a sitting posture: the **upper body**, which is located between the shoulders and the seat, and the **lower body**, located between the seat and the ground.

During push efforts, in addition to the push force, external forces include body weight and reactions at the support surface contacts. These reactions originate from the ground if the subject is standing, and from seat contact as well, if the subject is sitting. It is aimed to consider the role played by reactions at the support surface contacts, unlike the grip studies, which focused on local efforts on objects, and to consider the way in which the postural chain contributes to the motor act.

2. *The Transient Push Paradigm*

A transient push paradigm was considered in order to explore in greater detail how the postural chain contributes to the motor act.

This paradigm has been used in the past to study the control of motor responses under isometric conditions, in order to minimize several problems, which complicate experimental analysis. Rapid force impulses produced at a distal joint, like the elbow, were usually considered (see, for instance, Ghez and Gordon (1987), Gordon and Ghez (1987) or Corcos *et al.* 1990). In these studies, stops were used to prevent any body movement. In contrast, multi-joint pull and push tasks, performed by free-standing subjects, were chosen by authors like Whitney (1957) and Grieve (1979), or Cordo and Nashner (1982) and Lee and Patton (1997).

In this research, seated subjects were instructed to exert horizontal bilateral pushes on a bar, as rapidly as possible, up to their maximal force, and to maintain it for 5 seconds. They were asked to sit upright, and the apparatus was set to ensure their thighs were horizontal, their legs vertical, their upper limbs horizontally extended, and their hands gripping the bar. As the body was in contact with rigid surfaces (seat and footrests), making hand and foot movements impossible, the articulated body chain is said to be a **closed chain**. But the postural chain was not prevented from moving, as no additional support was used at the shoulder and trunk levels.

This paradigm offers many advantages: i) the muscular effort varies, but there is no movement of the extremity of the focal chain: as there are no hand movements, there are no "focal" kinematics; ii) since the subjects are in quasi-static conditions, the dynamics should be located in the postural chain (i.e. between the feet and the shoulders), which is divided into two parts: the upper and lower body; iii) since the subjects are seated, the mobility of the postural chain is easy to manipulate through a change in the ischio-femoral contact with the seat, without perturbing body balance. In this view, full ischio-femoral contact (100 BP, with BP for Bilateral Push) and a one-third contact (30 BP) were considered, the former being known to induce lesser lumbar spine and pelvis mobility than the latter.

2.1. BIOMECHANICAL MODELLING

A biomechanical model was elaborated in order to specify the role played by postural dynamic phenomena and to evaluate the effect of adherence at the contact level between the subject and the seat, as well as the footrests, in the course of transient efforts (Bouisset *et al.*, 2002).

To this end, the general equations of the mechanics were applied to the system. The subject's body was considered to be an isolated mechanical system. Consequently the forces applied to the system include the reaction forces originating from the body contact surfaces, in addition to body weight (Fig. 3).

In the Galilean coordinates system of the laboratory, the two dynamic scalar equations for the Centre of Gravity (CoG) movement in the sagittal plane are:

$$\begin{aligned} m\ddot{x}_G &= F_x + R_x \\ m\ddot{z}_G &= (R_z - W) + F_z \end{aligned} \tag{3}$$

In these equations, $\ddot{x}_G$, $\ddot{z}_G$ are the coordinates of CoG acceleration, W is the weight of the subject and m his/her mass; $-F_x$ and $-F_z$ are the antero-posterior and vertical external forces exerted by the bar on the subject (conversely, the forces exerted by the subject on the bar are equal to within the sign); R_x and R_z are the antero-posterior and vertical components of the reaction forces.

Furthermore, it can be written:

$$\begin{aligned} R_x &= R_{Sx} + R_{fx} \\ R_z &= R_{Sz} + R_{fz} \end{aligned} \tag{4}$$

where R_{Sx} and R_{fx} are the reaction forces along the antero-posterior axis at the seat and foot levels respectively, R_{Sz} and R_{fz}, the same reaction forces along the vertical axis.

The variation $\delta_y(G)$ of angular momentum (body angular acceleration times the moment of inertia) of this planar system is deduced from the moments of forces about the origin, O, of the laboratory reference frame as:

$$\delta_y(G) + m\ddot{x}_G z_G - m\ddot{z}_G x_G = x_G W - aF_z + hF_x - x_P R_z + bR_{Sx} \tag{5}$$

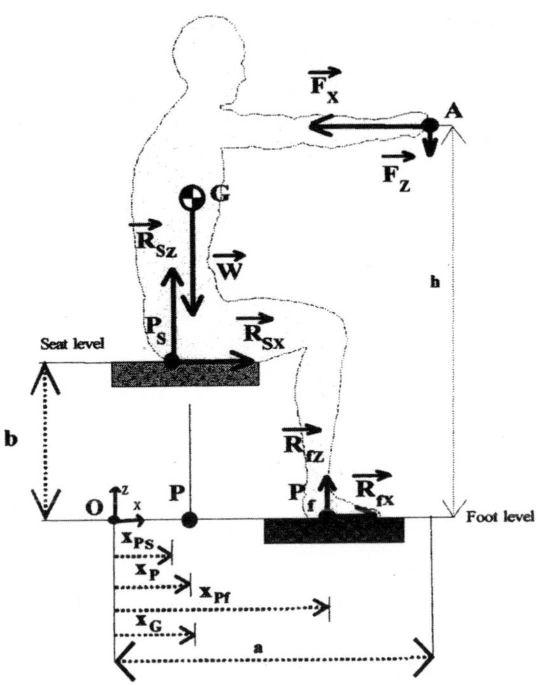

FIGURE 3. Biomechanical modelling. The diagram of external force vectors corresponds to a two-handed push exerted on a bar by a seated subject. Horizontal and vertical reaction forces, F_x and F_z, exerted *on* the subject; R_{Sx} and R_{Sz}, R_{fx} and R_{fz}: antero-posterior and vertical reaction forces at the seat and foot levels; W: is the weight of the subject, acting through the CoG line; x_P, x_G: x coordinates of CoP (P) and CoG (G) according to origin O; h: vertical distance from the dynamometric bar (A) to the footrest plane; a: horizontal distance of the bar to O; b: vertical distance between seat and foot levels.

The quantities a, b and h are parameters of the experimental set-up which are adjustable according to the subject's anthropometrical data (Fig. 3). The x-coordinates of the Centre of Pressure (CoP) at the seat and feet are denoted respectively as x_{PS} and x_{Pf}. The x-coordinate x_P of the global CoP is given by:

$$x_P = x_{PS}\frac{R_{Sz}}{R_z} + x_{Pf}\frac{R_{fz}}{R_z} \qquad (6)$$

At the end of the push effort, a new mechanical equilibrium occurs, and the equations of balance can be deduced from (3) and (5), that is:

$$\begin{aligned} R_x &= -F_x \\ R_z - W &= -F_z \end{aligned} \qquad (7)$$

and:

$$\begin{aligned} x_G - x_P = {} & [(R_z - W)/W](x_P - a) + (R_{Sx}/W) \\ & \times (h - b)h + (R_{fx}/W)h \end{aligned} \qquad (8)$$

This equation becomes simpler if R_{fz} and R_{fx} are negligible in comparison to R_{Sz} and R_{Sx} respectively, which will be proven later (section 2.2.2.2):

$$x_G - x_P = [(R_{Sz} - W)/W](x_P - a) + (R_{Sx}/W)(h - b) \qquad (9)$$

Equation (9) can be rearranged in order to get push force:

$$\begin{aligned} -F_x = {} & (R_{Sz} - W)(a - x_P)/(h - b) \\ & + W(X_G - X_P)/(h - b) \end{aligned} \qquad (10)$$

Hence, $-F_x$ increases as a function of x_P, x_G and R_{Sz}. In particular, if x_G is negligible, and $(R_{Sz} - W)$ is constant at the end of push effort, $-F_x$ is proportional to the CoP backward displacement.

Furthermore, equation (9) can be rewritten, taking the adherence ratio into account:

$$\begin{aligned} (x_P - x_G) + {} & [(R_{Sz} - W)/W](x_P - a) \\ & + (\mu_S R_{Sz}/W)(h - b) = 0 \end{aligned} \qquad (11)$$

Equation (11) relates CoP displacement to vertical reaction forces and adherence ratio (μ_S, at the seat level). However, it does not result in a cause and effect relation between these three factors.

An experimental protocol was designed to measure the various terms of the model (Bouisset *et al.*, 2002). To this end, the subjects were seated on a custom-designed device (Lino, 1995). Three rectangular force plates, linked by a rigid frame, measured reaction forces and positions of the centre of pressure at the foot and seat levels. The CoG coordinates (x_G, z_G), along the antero-posterior and vertical axes, were deduced from the CoG acceleration (equations (3) by a double integration. As the x origin was taken at the global CoP at rest, the x coordinates measured the x displacement. Force transducers measured the antero-posterior and vertical forces exerted by the bar on the subject, and conversely.

2.2. TRANSIENT PUSH INDUCES POSTURAL DYNAMICS

2.2.1. Transient Push Force. As the subjects were asked to push horizontally, the horizontal external

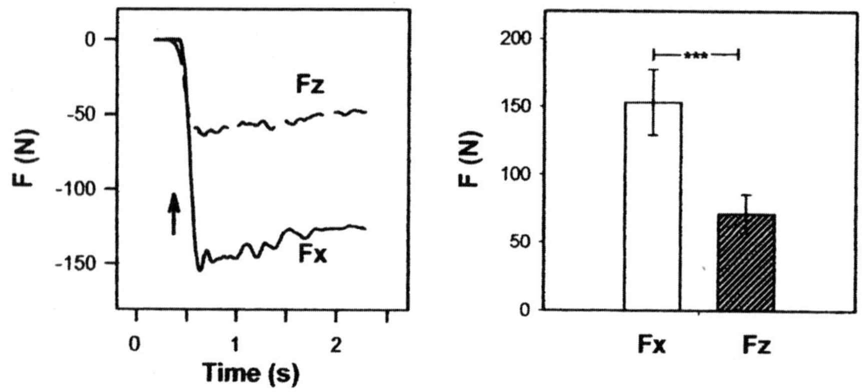

FIGURE 4. Transient push force. Left: F_x and F_z refer to the antero-posterior and vertical forces applied by the bar *on* the subject. Mean curve calculated over seven trials performed by the same subject. The arrow indicates the onset of push force. Right: Absolute peak force values. Means and standard deviations were calculated over seven subjects. ***: $p < 0.001$ (highly significant).

force (F_x) developed during the transient push effort was a measure of the intended act. It could be observed that the corresponding force exerted by the bar on the subject was negative (Fig. 4). Also, negative vertical (F_z) forces were developed during the transient push effort. According to the sign conventions, the push effort on the bar was directed upwards, as well as forwards. Both components, F_x and F_z, increased progressively. They displayed the same well-known force-time shape (Wilkie, 1950) when they are plotted against time, and the horizontal force, F_x, peaked at a mean value which is almost double $F_z(-153 + /-24$ N, as compared to $-71 + /-14$ N). The F_z vs. F_x relationship was exponential ($F_z = a(1 - e^{-bF_x})$).

Thus, task achievement included two force components. One, the horizontal force, measured the task **performance**, while the other, the vertical force, appeared to be a "by product" of the motor act. In accordance with previous studies (for a review, see Bouisset and Le Bozec, 2002), F_x which is an input of the motor system, can be considered to provoke a **perturbation** of body balance.

2.2.2. Body Dynamics. The equations resulting from the biomechanical model included global quantities, as well as local ones, measured at the seat and foot levels. Their time variations yielded during the push effort were considered and evaluated from experimental data.

2.2.2.1. GLOBAL DYNAMICS. The time course of the resultant reactions originating from the supports (R_x, R_z along the antero-posterior and vertical axes), that of CoP and CoG displacements (x_P and x_G along the antero-posterior axis), as well as the reaction (F_x) to the horizontal push force ($-F_x$), are displayed in Fig. 5.

All the time courses show the same sigmoid profile, to within the sign. It can be observed that when the push force increased, the reactions originating from the supports, R_x and R_z, increased as well, that is the subject exerted downward and backward efforts on the supports. It was also observed that F_x (and R_x), as well as F_z (and R_z), displayed opposite signs, in agreement with the action-reaction law (equation (3)): the perturbation applied on the body at the hand level was instantaneously counter-acted by the reactions at the seat and foot levels. Simultaneously, x_P decreased, showing that the CoP moved backward. More precisely, CoP unlike CoG displacement was found to be great: (-108 $+/-94$ mm as compared to $-5 + /-2$ mm), and $x_P - x_G$ decreased progressively, showing that CoP withdrew from CoG.

It was also observed that the onsets of R_x ($-60 + /$ -5 ms), R_z ($-60 + /-7$ ms) and x_P ($-62 + /-6$ ms) preceded highly significantly ($p < 0.001$) the onset of the push force increase. In other words, there were Anticipatory Postural Adjustments (APAs).

In addition, parametric relations were considered (R_x vs. F_x, R_z vs. F_z, x_P vs. F_x and $\mu = R_x/R_z$ vs. F_x). The relationship between F_x and R_x established that R_x was approximately proportional to F_x (Fig. 6), as was the relationship between R_z and F_z. However, systematic, though minor, deviations from the bisector line were observed. In accordance with the

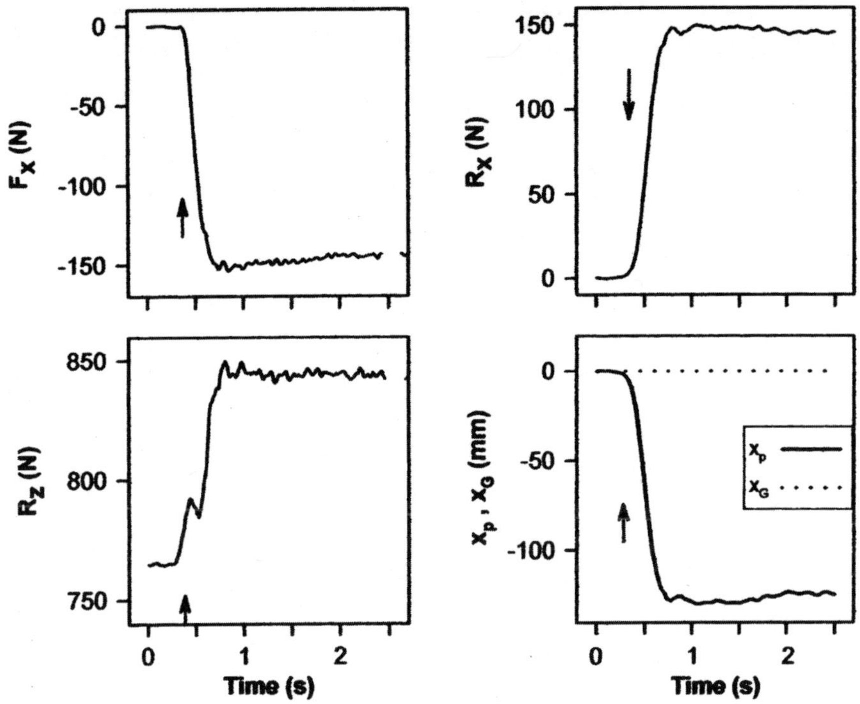

FIGURE 5. Push force, global reaction forces and centre of pressure time courses. 1st row: Left column: horizontal reaction to push force (F_x); right column: global reaction forces at the seat and foot contacts (R_x along the antero-posterior axis). 2nd row: Left column: global reaction forces at the seat and foot contacts (R_z along the vertical axis); right column: global CoP displacements (x_P and x_G along the antero-posterior axis). The vertical arrow indicates the onset of push force Mean curve calculated over seven trials performed by the same subject.

equations (3), these discrepancies between the actual profile and the linear one result from **inertial forces**, that is, body link acceleration (inertial forces = subject's mass times CoG acceleration). The same result was obtained when x_P was plotted against F_x. The discrepancy from linearity could also be attributed to inertial force effects, that is, angular momentum variations in this instance, according to equation (5). In other words, inertial forces flowing throughout the body chain underlie dynamic phenomena. More specifically, it can be said that the articulated body chain was in a state of *dynamic* equilibrium.

Moreover, the subject was in a fixed posture, with his upper limbs outstretched and his hands grasping the bar. Therefore, body link accelerations could only originate from the rest of the body, that is, from the **postural chain** (Le Bozec *et al.*, 1997). The role played by the postural chain was confirmed by the backward displacement of the centre of pressure, corresponding to hip extension. This displacement requires a modification in the distribution of reaction forces between the body and its supports, which local biomechanics help to specify.

2.2.2.2. LOCAL DYNAMICS. The partitive dynamic method allows a more precise statement of the postural counter-perturbation. To this end, local dynamics were assessed, that is reaction forces and CoP positions at the seat and foot levels (Fig. 7).

It appeared that the local curves displayed the same sigmoid profile as the global ones. However, this similarity held true only to within the sign. Indeed, the vertical force variations at the seat and foot levels (R_{Sz} and R_{fz}) yielded opposite signs (Fig. 7, middle row), unlike antero-posterior variations at the same levels (R_{Sx} and R_{fx}).

More precisely, vertical reaction forces increased at the seat level, whereas they decreased at the foot level, that is, the upper body was pushing on the seat during the transient push effort. In other words, there is a transfer of the reaction forces to the seat support up to the end of push, resulting in progressive anchoring of the upper body to the seat. The vertical foot

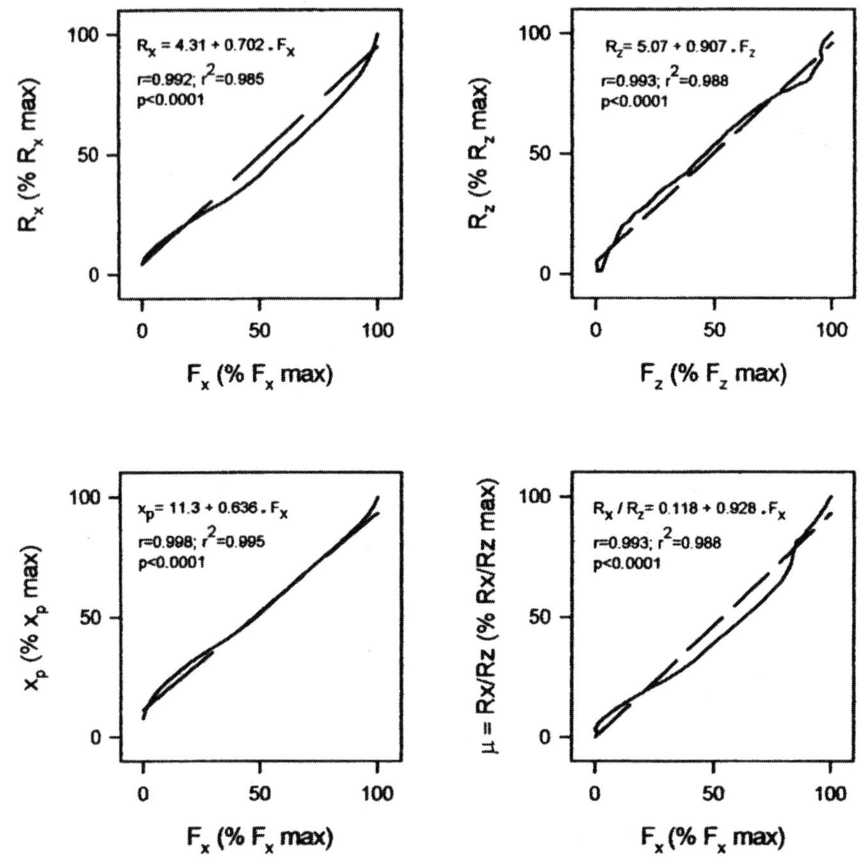

FIGURE 6. Parametric relationships during maximal ramp pushes. The regression lines are represented as a broken line; r: Bravais-Pearson coefficient of correlation. Mean curves calculated over seven trials performed by the same subject. 1st row: Left column: R_x (global reaction forces along the antero-posterior axis) plotted against horizontal push force (F_x); right column: R_z (global reaction forces along the vertical axis) plotted against vertical push force (F_z) 2nd row: Left column: x_P (global CoP displacement along the antero-posterior axis) plotted against horizontal push force (F_x); right column: μ (adherence ratio) plotted against horizontal push force (F_x). All the quantities are expressed as a percentage of their maximal value.

reactions favour forward body destabilization, and also contribute to CoG antero-posterior acceleration. In addition, because they yield an opposite sign to the upper body vertical reactions, lower limb dynamics contribute to upper body vertical force production and favour pelvis rotation.

Consequently, the increase in upper body vertical reaction forces and the decrease in lower body forces reinforce the ability to counteract the perturbation induced by the push effort, that is, it enhances **Posturo-Kinetic Capacity** (PKC) (Bouisset and Zattara, 1983; Bouisset *et al.*, 2002). In other words, there was a co-ordinated action of the upper and lower body.

In addition, R_{Sx} and R_{Sz} peak values were highly significantly greater than the R_{fx} and R_{fz} peak values. Also, the global reaction forces (R_x and R_z) were nearly equal to the reactions at the seat contact surface (R_{Sx} and R_{Sz}). In particular, R_{Sz} was almost equal to R_z. Hence the CoP backward displacement at the end of transient effort, x_P, (−108+/−30 mm) was very close to x_{PS} (−94+/−17 mm). As a consequence, it appears that the push effort entails a transfer of the global CoP to the upper body CoP.

It was also observed that the onsets of R_{fx} (−64+/−4 ms), R_{fz} (−61+/−7 ms) and x_{Pf} (−64+/−2 ms) preceded very significantly

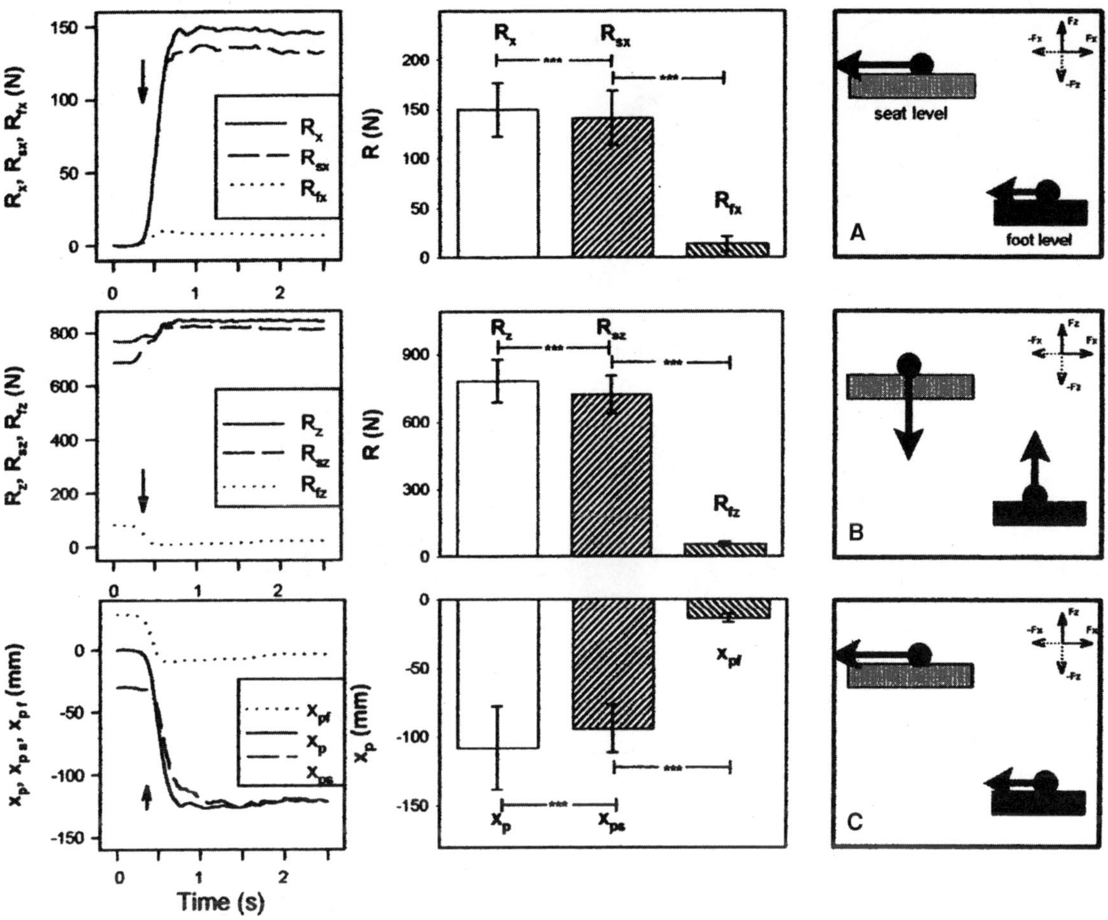

FIGURE 7. Local reaction forces and centre of pressure time courses. Left column: Reaction forces and centre of pressure time courses. From top to bottom: global reaction forces (R_x along the antero-posterior axis) and local reaction forces at the seat (R_{Sx}) and foot (R_{fx}) levels; global reaction forces (R_z) and local reaction forces at the seat (R_{Sz}) and foot (R_{fz}) levels along the vertical axis; global and local centres of pressure along the antero-posterior axis (x_P, x_{PS} and x_{Pf})). Mean curve calculated over seven trials performed by the same subject. Middle column: Peak values for the same reaction forces and centres of pressure. Means and standard deviations were calculated for all seven subjects. ***: $p < 0.001$ (highly significant). Right column A and B: Direction of the efforts exerted by the subject on the seat and footrests; C: Displacement of the centres of pressure at the seat and foot levels.

($p < 0.01$) the push force increase. They also preceded very significantly the onset of $R_{Sx}(-61+/-5$ ms), $R_{Sz}(-60+/-7$ ms) and $x_{PS}(-62+/-2$ ms). Therefore, there were **Anticipatory Postural Adjustments** (APAs), and the APA sequence started at the foot level. Moreover, there was no significant difference between the onsets of R_{Sx}, R_{Sz} and x_{PS}.

To summarise: i) upper and lower body actions are coordinated; ii) upper body dynamics appear to play a major role in postural stabilization; iii) APAs proceed according to a bottom-up sequence.

3. Postural Chain Mobility, A Key Factor for Performance

This paradigm made it possible to manipulate the surface contacts between the body and its physical supports. For instance, it was easy to reduce the ischio-femoral contact with the seat, from complete contact (100 BP) to one-third contact (30 BP), without perturbing balance. Indeed, this modification did not change the overall support contour: the support base perimeter remained the same (Fig. 8).

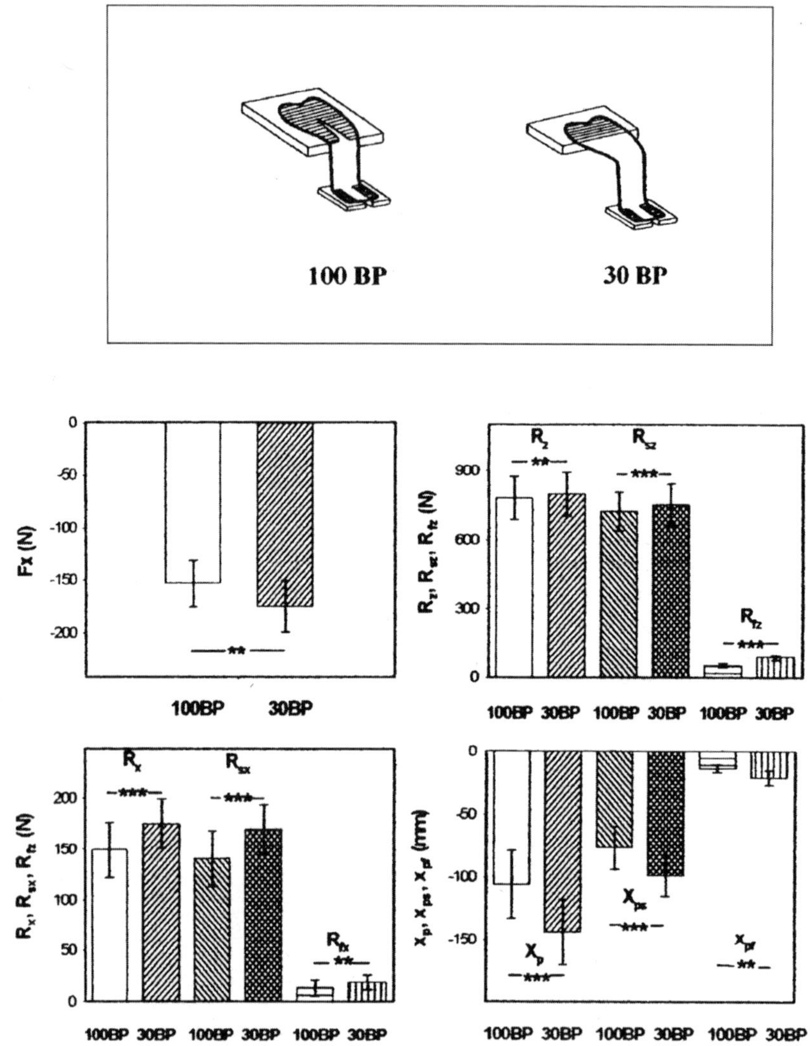

FIGURE 8. Influence of postural chain mobility on biomechanical variables. Top inset : Schematic representation of complete and one-third ischio-femoral contacts. 100 BP: complete ischio-femoral contact; 30 BP: one-third ischio-femoral contact. Bottom : Peak values Left column: Horizontal push forces (F_x) exerted on the subject (first row); antero-posterior global (R_x) and local reaction forces at the seat (R_{Sx}) and foot (R_{fx}) levels (second row). Right column: vertical global (R_z) and local reaction forces at the seat (R_{Sz}) and foot (R_{fz}) levels (first row); global (x_p) and local centres of pressure along the antero-posterior axis (x_{PS}, x_{Pf}) (second row). Means and standard deviations were calculated for all seven subjects. 100 BP: complete ischio-femoral contact; 30 BP: one-third ischio-femoral contact **: $p < 0.01$ (very significant); ***: $p < 0.001$ (highly significant).

When ischio-femoral contact was reduced, the peak push force, F_x, was very significantly increased (Fig. 8). This might appear surprising, though only at first glance. Indeed, performance enhancement was associated with significant dynamics increases, which were proven by the maximal values of the global biomechanical variables under consideration (R_x, R_z and x_P). The local biomechanical variables at the seat and foot level (R_{Sz}, R_{fz}; R_{Sx}, R_{fx}; x_{PS} and x_{Pf}) were also increased and yielded the same general features, when ischio-femoral contact was limited (Fig. 8).

It is known that **pelvis mobility** is modified by a reduction in the seat contact area from 100 BP to 30 BP. Indeed, when ischio-femoral contact is limited, such as in the 30 BP posture, the pelvis can rotate with respect to the seat about an axis passing through the contact of the ischiatic tuberosities with the seat, and, with respect to the thighs about an axis passing through the femoral heads (Vandervael, 1956). On the other hand, when ischio-femoral contact is complete, that is in the 100 BP posture, the thighs are in close contact with the seat and cannot be displaced: the pelvis can only move about an axis passing through the femoral heads. Therefore, pelvis mobility is less in the 100 BP than in 30 BP condition. As a consequence, CoP displacement is greater in the 30 BP condition.

Moreover, according to the PKC theory (Bouisset and Zattara, 1983; Bouisset and Le Bozec, 2002), if movement induces a dynamic perturbation, the counter-perturbation must be dynamic as well. Now, given that transient push efforts induce dynamics, the postural counter-perturbation must also be dynamic, in order to attain the intended performance. Consequently, if postural chain mobility is constrained in one way or another, fewer postural segments can be accelerated, counter-perturbation is limited, and performance reduced. In other words, the increased mobility of the postural chain favours postural dynamics, and hence PKC, which produces greater force at the end of the effort.

These results generalize to ramp efforts those obtained by Lino *et al.* (1992) for pointing movements performed under the same two support conditions. When ischio-femoral contact is reduced, performance (that is, maximal velocity, in the pointing movement) increases significantly, in parallel to dynamic postural phenomena. Thus, it does not matter whether the effort is, according to the physiological terminology, "dynamic" as in the Lino *et al.* (1992) study, or "static" (but "anisotonic") as in this one. In both conditions, the perturbing effect on balance is associated with a variation of muscular force. When the contact area is reduced, that is, when postural chain mobility is greater, performance is enhanced. In terms of biomechanics, it can be said that transient efforts are necessary for the body system to proceed from the initial to the final mechanical equilibrium, which has been already defined (equations (7) and (8)).

In conclusion, postural compensation to the perturbation provoked by an effort depends not only on the support base perimeter, that is the **stability area**, but also on postural chain mobility, that is on the **free play** of postural joints. In this study, it is a function of pelvis and lumbar column mobility. As a consequence, postural chain mobility appears to be a key factor in PKC.

4. *Global and Local Adherence Ratios*

The adherence ratio has been defined as "friction use" (see section 1–2). It reflects how the CNS takes into account the contact forces between the body and its physical environment in order to perform the motor act efficiently. In addition, by this very fact, AR corresponds to the ratio of tangential to normal reaction forces at the contact surfaces, and consequently to the actual angle of adherence (Fig. 1). Adherence ratios were considered globally, that is, in a whole, or locally, that is, at the foot and seat surface contacts.

4.1. TRANSIENT PUSH INDUCES A CONTINUOUS INCREASE IN FRICTION USE

In the earliest instants of push, the global AR ($\mu = R_x/R_z$) was almost nil, and then increased sharply, up to the peak value displayed at the end of push (Fig. 9, first row): there was a continuous increase in AR, that is AR got closer and closer to CoF. Similar results were found when the local ARs at the seat ($\mu_S = R_{Sx}/R_{Sz}$) and the foot ($\mu_f = R_{fx}/R_{fz}$) supports were considered (Fig. 9, second and third rows).

In addition, the global peak Adherence Ratios (pAR) were highly significantly greater when the ischio-femoral contact with the seat was changed, from complete (0.18 +/− 0.03) to one-third (0.21 +/− 0.02) contact (Fig. 10). The increase was related to increases in reaction forces at the seat and foot levels (Fig. 8). Therefore, there was increased "friction use" when postural chain mobility was enhanced. Similar results were reported in the study of pointing tasks in the same postural conditions (Lino, 1995).

Local pARs yielded the same feature. Indeed, the pAR values at the seat support were also highly significantly higher for 30 BP (0.20 +/− 0.02) than for 100 BP (0.17 +/− 0.03) (Fig. 10). These values were lower than the coefficient of friction (0.25), which was measured directly at the seat (and footrests) fabric-wood interface. Therefore, a safety margin can be assumed, in accordance with Johansson and Westling (1984). On the other hand, the pAR values at the foot (0.29 +/− 0.06 for 30 BP and 0.23 +/− 0.11 for 100 BP) were not significantly different (Fig. 10), and were so close to the CoF, that slipping cannot

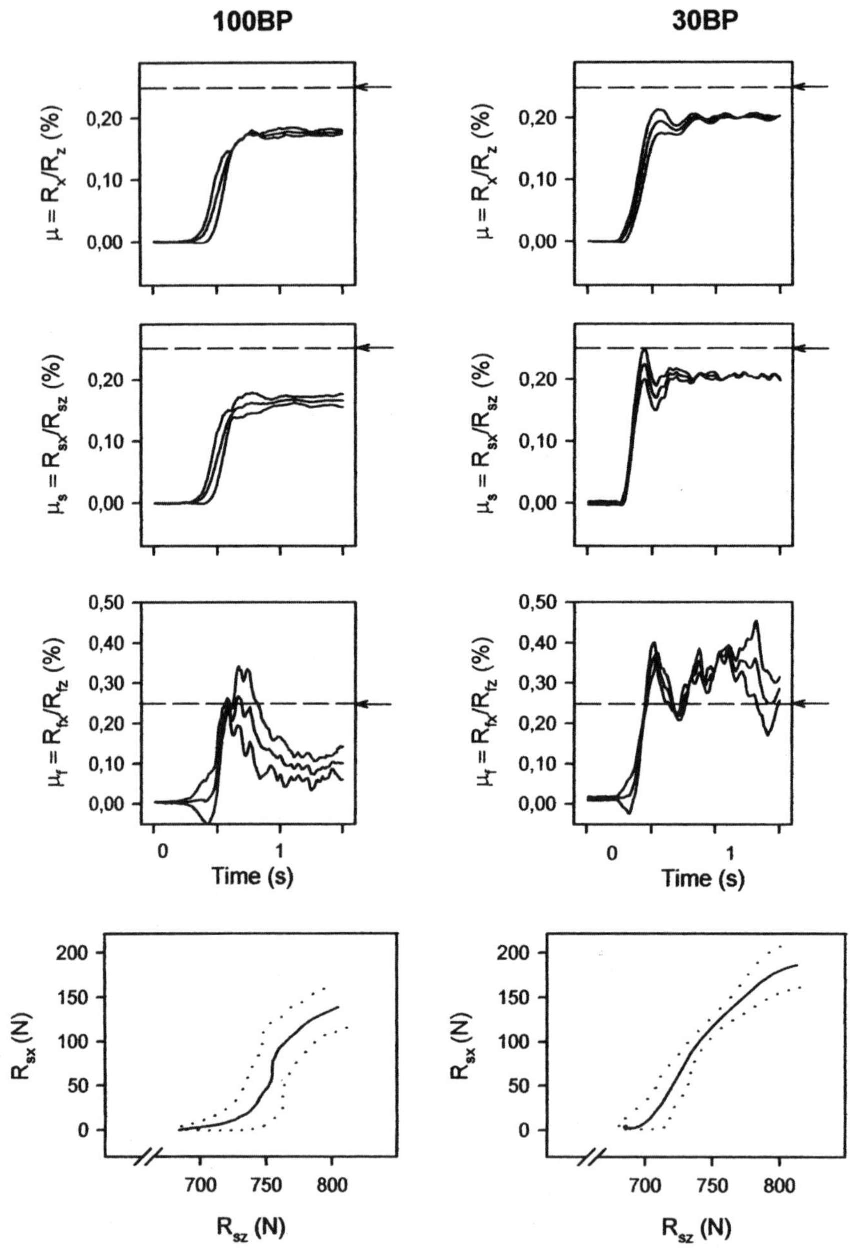

FIGURE 9. Instantaneous adherence ratio variations. 1st row: Global adherence ratio ($\mu = R_x/R_z$, in %) as a function of time. 2nd row: adherence ratio at the seat contact surface ($\mu_S = R_{Sx}/R_{Sz}$ in %) as a function of time. 3rd row: adherence ratio at the foot contact surface ($\mu_f = R_{fx}/R_{fz}$, in %) as a function of time. 4th row: tangential component at the seat contact surface (R_{Sx}) plotted against the corresponding normal component (R_{Sz}). Mean +/− one standard deviation was calculated over seven trials performed by the same subject. 100 BP: complete ischio-femoral contact; 30 BP: one third ischio-femoral contact The hatched line indicated by an arrow corresponds to the CoF value (0.25).

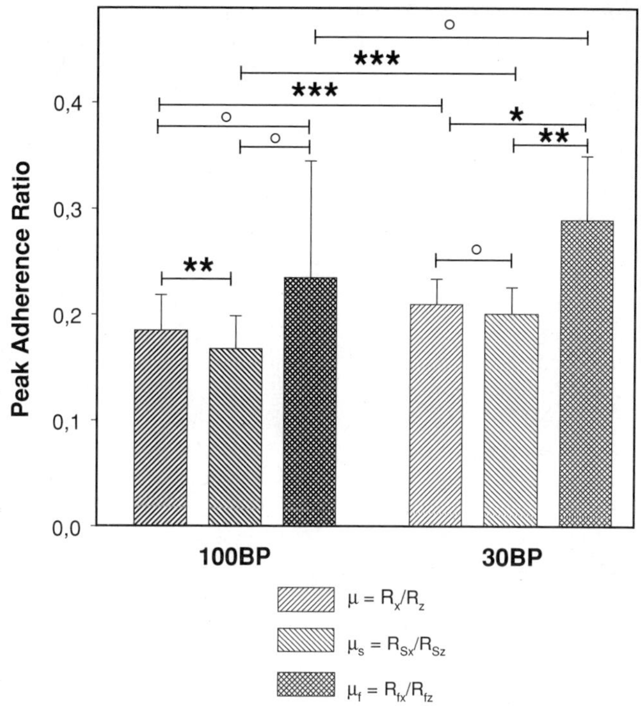

FIGURE 10. Global and local peak adherence ratios. Global ($\mu = R_x/R_z$) and local ($\mu_S = R_{Sx}/R_{Sz}$ and $\mu_f = R_{fx}/R_{fz}$) adherence ratios at the seat (subscribe S) and foot (subscribe f) levels. The coefficient of friction between the subjects and the seat (and the footrests) was 0.25. Means and standard deviations were calculated for all seven subjects. 100 BP: complete ischio-femoral contact; 30 BP: one-third ischio-femoral contact. °: $p > 0.05$ (non significant); *: $p < 0.05$ (significant); **: $p < 0.01$ (very significant); ***: $p < 0.001$ (highly significant).

be excluded at the very end of the push effort, at least for some subjects, as exemplified in Fig. 9 (third row).

Consequently, the data obtained at the foot support might suggest that the safety margin would be respected only under certain conditions. Indeed, such possibilities could occur when the orders regarding posture were not compatible with the intended task-movement performance and/or with body stability. One can wonder whether this is not the case in these experiments, as a lack of contact with the footrests was shown to favour maximal push force (Gaughran and Dempster, 1956). Moreover, slipping at the foot level might not be a problem: given that the subject was holding the bar, global posture was not insecure. Therefore, local ARs could be supposed to be managed with reference to the CoF, but in a different way according to the intended performance and the effect of local slipping on body stability.

4.2. ADHERENCE RATIO INCREASE RESULTS FROM SIMULTANEOUS INCREASE OF REACTION FORCE COMPONENTS

According to equation (2), AR is the ratio of R_x to R_z, that is, of the instantaneous horizontal to the vertical reactions at the contact surfaces. Consequently, an increase in AR could result from simultaneous or independent variations in R_x and R_z. Simultaneous variations have been reported in various tasks, such as walking (Strandberg, 1983), prehension (Johansson and Westling, 1984) and pointing (Lino, 1995). Results on global and local body dynamics (see sections 2.2.2 and 2.2.3, as well as Fig. 5 and 7) were in favour of such an assumption.

In order to deepen the question, the instantaneous variations of the reaction forces at the seat level (R_{Sx} and R_{Sz}) were considered, given the major role devoted to the reaction forces at the seat, and consequently to upper body dynamics (section 2.2.2.2; Fig 7 and

Fig. 8). Indeed, global reaction forces (R_x and R_z) were found to be nearly equal to the reactions at the seat contact surface (R_{Sx} and R_{Sz}). In addition, the global pAR was found to be approximately one percent greater than the local pAR at the seat support for both ischio-femoral contact conditions (0.18 as compared to 0.17 for 100 BP; 0.21 as compared to 0.20 for 30 BP).

At the beginning of push, the vertical component R_{Sz} increased faster than the horizontal component R_{Sx} (Fig. 9, fourth row). After the inflexion point, both R_{Sx} and R_{Sz} continued to increase, but the slope (dR_{Sx}/dR_{Sz}) decreased. Subsequently, the R_{Sx} increase was greater than the R_{Sz} one. Hence, the vertical and horizontal force components displayed *simultaneous* increases. Therefore, the continuous increase in AR originated mainly from a simultaneous and continuous increase of the reaction forces at the seat (R_{Sx} and R_{Sz}) during the push effort (Fig. 7). Similar results have been reported for pointing tasks by Lino (1995), suggesting that it is not a particular feature. More generally, as simultaneous vertical and horizontal reaction forces were observed, it can be surmised that the stabilizing reactions imply that they are modulated in such a way that the maximal push force is developed with the aim of preventing slipping at the end of push, that is under the guidance of AR.

Lastly, it is interesting to keep in mind that pAR, as well as maximal external force, were enhanced when the coefficient of friction was increased (Kroemer, 1974; Grieve, 1979; Gaudez *et al.*, 2003). Therefore, in order to enhance pAR and maximal force, there are two possibilities: increase postural chain mobility (see section 3) and/or increase the CoF at the support surfaces. In other words, the **Maximal Voluntary Force** (MVF) does not depend only on the prime movens maximal force, that is of those muscles that are primarily responsible for the intended movement. MVF is also limited by the CoF value, which in turn limits the AR maximal value, to within a possible security margin. For a given CoF, it also depends on postural conditions, such as the mobility of the postural chain and support base perimeter. In other words, postural factors limit the maximal effort that the muscles can exert: the capacity to oppose the perturbation provoked by the voluntary effort, that is Posturo-Kinetic Capacity, modulates the intensity of the voluntary effort in order to prevent slipping.

To summarize: i) continuous global as well as local AR increases were observed in the course of the push effort up to values which were close to CoF; ii) the vertical and the horizontal reaction forces yielded simultaneous increases; iii) the risk of slipping on the supports during the effort was bounded by the postural chain's capacity to afford convenient AR values, insofar as adherence is required to make the push effort possible; iv) R_x variations are assumed to be modulated under AR control, that is, in such a way as to prevent slipping at the end of the push.

5. *Postural Control and Adherence*

It is well known that there are many ways to approach motor control, and that the complexity of the process leads to some speculations. This experimental approach provides new data of a biomechanical order. It is interesting to examine how they help clarify certain aspects of motor control, and in particular the adherence effects on motor programming.

The biomechanical data allowed a description of the motor sequence, taking place between initial and final static equilibrium. They establish that a continuous increase of body dynamics is associated with the continuous increase of the push effort: the postural chain is in a state of dynamic equilibrium. Body dynamics originate at the footrest level and proceed up to the hand level, according to a bottom-up sequence. A continuous dynamic increase at ischio-femoral contact is associated with rear pelvis rotation and CoP displacement. In this process, upper body dynamics appear to play a major, though not exclusive, role for postural stabilization during the effort. Postural stabilization depends on postural chain mobility, that is, on the free play of postural joints (pelvis and lower spine mainly, in these conditions). The adherence ratio increases continuously during the effort, up to a value, which appears to correspond to the coefficient of friction to within a safety margin, at least at the seat level.

5.1. RATE OF FORCE RISE, AS A PLANNED VARIABLE

As reviewed by Macpherson (1991), several authors have proposed that motor act parameters are controlled hierarchically. The higher-level parameters could be assumed to be global, usually mechanically defined, and related to the goals of the movement. They would participate in determining the values of the more local lower level variables in any given solution of a motor problem.

According to Bernstein (1935; Amer. Translation, 1967), a motor task evolves a voluntary movement, and is planned in terms of kinematics in the external Cartesian space, that is, in the task space. In other words, the goal of the planned movement is expressed in terms of its path, that is, the displacement of the tip of the distal segment (usually called "end-point" or "working point"). To this end, the system should be able to perform an internal simulation of a planned movement, where its actual parameters are taken into account. Then, the commands would lead to changes in the activation of the muscles controlling the joints mobilized by the voluntary movement. While this viewpoint has been widely adopted, the authors differ as to the relative role devolved to spinal reflexes and central command (see Latash 1993 for a detailed review).

In this study, there were no "focal" kinematics. Hence, the possible internal simulation could not follow any relation between the end-point kinematic variables. Consequently, one might envision a relation between some of the variables characterizing the external force exerted at the end-point, which could be considered as the planned variable. As isometric forces are developed as quickly as possible up to the maximum, the parameter of the planned motor act would be the **rate of force rise**, in accordance with Gordon and Ghez (1987). Indeed, these authors have shown that peak isometric force is achieved by a proportional modulation of the rate of force rise, which has been confirmed by Corcos *et al.* (1990). Parallel variations of the peak force and the rate of force rise were also found in transient push efforts (Le Bozec and Bouisset, 2004).

Even if the postural chain were free to move in these experiments, contrary to the single-joint paradigms considered by Gordon and Ghez (1987) and Corcos *et al.* (1990), there is no reason to exclude that the motor act is planned in terms of rate of force rise.

5.2. CENTRE OF PRESSURE, AS A POSTURAL CONTROL VARIABLE

However, the role of the postural chain cannot be ignored. Indeed, it has also been proposed by Gelfand *et al.* (1966), revisiting Bernstein's ideas (1935), that motor tasks include a focal and a postural component, one referring to the body segments that are mobilized in order to perform voluntary movement directly, and the other, to the rest of the body which is involved in the stabilizing reactions. These definitions suggest that the two parts must be conceived as functional. They transcend simple anatomical partitioning, and are assumed to cope in motor control. In this context, the possibility of postural control is justified.

Various postural control variables have been proposed in literature, mainly CoG, CoP and Rx, which were assumed to be at a lower hierarchical level. Several authors have suggested that CoG and CoP are postural control variables for postural tasks (for a review, see Horak and MacPherson, 1995). The role of one or the other is still under discussion (Lacquaniti and Maioli, 1994), and it is very likely that it will depend on the task conditions. On the other hand, authors have claimed that the contact forces at the feet, namely the tangential ones, are high-order control parameters, at least for quadruped posture (MacPherson, 1988, 1991).

In the biomechanical model, which has been proposed above (equations (3) and (5)), five main quantities appear to be involved in push efforts (F_z, R_x, R_z; x_P and x_G). These quantities can be *a priori* identified as control variables, given that the horizontal push force, F_x, is the planned variable. They are linked by the three independent equations (3) and (5). In order to limit the risk of slipping, there is a complementary inequation, which expresses the no slipping condition $\mu \leq \mu^*$ (relation (2)).

In these experiments, the CoG displacement was found to be negligible in contrast to CoP displacement (Fig. 5). However, the postural constraints to which the subjects have to comply limit the number and amplitude of the anatomical degrees of freedom, suggesting that the CoG displacement might be only very limited. Therefore, CoP displacement appears to be a better candidate than CoG as a postural control variable. In addition, the only possibility for the postural chain to develop a counter-perturbation to the balance perturbing push force (and consequently to exert a significant push force), originates from a CoP displacement, in accordance with the comments on equation (10). Such a contention is reinforced by the EMG data: the activation of the pelvis extensor muscles (Gluteus Maximus in cooperation with Biceps Femoris), which provokes pelvis backward rotation, is in relation with CoP rear displacement (Le Bozec *et al.*, 2001; Le Bozec and Bouisset, 2004). In this context, it is interesting to observe that transient push force and CoP displacement presented the same sigmoid time course profile (Fig. 5), and that their relationship was approximately linear (Fig. 6), which could simplify the command.

Once it is admitted that CoP is a postural control variable, the question is to determine the role induced by Newton's law and the no slipping condition on the other three biomechanical quantities (R_x, R_z and F_z). It has been shown that CoP rear displacement results from a coordinated action of the lower and upper parts of the postural chain. A rough outline of the question points to the major role played by the pelvis, that is, to rear pelvis rotation. Such a rotation has been shown to induce an increase in R_x (Fig. 5), and primarily an increase in R_{Sx}, that is, at the seat contact surface (Fig. 7). This increase constitutes a necessary counter-perturbation to the destabilizing horizontal push force, F_x, according to equation (3).

Moreover, pelvis rotation is associated with an increase in R_z (mainly R_{Sz}), whose destabilizing effect would be compensated by F_z. In addition, AR is kept under the CoF (mainly μ_S, at the seat level), taking the safety margin into account (Fig. 9). This suggests that there is a pairing of the horizontal and vertical reaction forces, in order to prevent slipping at the end of push.

If it is admitted that the R_{Sx} increase is the result of pelvis rotation, and that the simultaneous R_{Sz} increase is a biomechanical consequence of this rotation, the μ_S value could be one of the rules for controlling CoP displacement (equation 11). Hence, the actual coefficient of friction value might be implemented in the motor program, as it has generally been supposed since Westling and Johansson (1984). In this context, it is interesting to observe that the relationship between global AR and transient push force (and consequently CoP displacement) was approximately linear (Fig. 6), which could simplify the command.

Finally, the stabilizing reactions are actuated in order to integrate sensory information originating in the body contact surfaces. The forces exerted on these surfaces are assumed to be calibrated so as to respect the adherence limit. Of the information taken into account, it is generally considered that **haptic** information plays a major role (see Wing *et al.*, 1996, for a review). Unfortunately, there are presently very little, if any, physiological data on ischio-femoral afferent haptic signals. For the successful elaboration of a motor task, CNS control processes may use feed-forward mechanisms, which are based on internal models, that not only program the action, but also predict deviations induced by perturbations, and appropriate responses to restore the initial plan (Ghez *et al.*, 1995). The APAs, which were reported in this study, confirm feed-forward postural control. But they do not make it possible to settle in favour of two parallel controls responsible for the intended task movement and related balance stabilization (Alexandrov *et al.*, 2001), or a single control process for a whole-body movement, leading to these two distinct peripheral patterns classified as focal and postural (Latash, 1993; Aruin and Latash, 1995).

To summarize, in the context of a hierarchical organization, it could be proposed that the planned variable of the isometric transient effort is the rate of the voluntary force increase. The displacement of CoP is suggested as being the postural control variable, which is at a lower hierarchical level. The limit of adherence ratio, with reference to the coefficient of friction, could be one of the rules for controlling CoP displacement and muscle activation in order to accomplish voluntary ramp effort.

References

Alexandrov AV, Frolov AA, Massion J (2001) Biomechanical analysis of movement strategies in human forward trunk bending. I. Modeling. Biol Cybern., 84 : 425–434.

André-Thomas (1940) Equilibre et équilibration. Paris: Masson et Cie., 1–568.

Aruin AS, Latash ML (1995a) The role of motor action in anticipatory postural adjustments studied with self-induced and externally triggered perturbations. Exp Brain Res 106: 291–300.

Aruin AS, Latash ML (1995b) Directional specificity of postural muscles in feed-forward postural reactions during fast voluntary arm movements. Exp Brain Res 103: 323–332.

Bernstein N (1967) The Co-ordination and regulation of Movements. Oxford, UK: Pergamon, 1–196.

Bouisset S and Le Bozec S (2002) Posturo-kinetic capacity and postural function in voluntary movements. In Latash, ML (Ed): Progress in Motor Control, volume II: Structure-Function Relations in Voluntary Movements. Human Kinetics. Chapter 3 : 25–52.

Bouisset S, Le Bozec S, and Ribreau C, (2002) Postural dynamics in maximal isometric ramp efforts. Biol. Cybern. 87: 211–219.

Bouisset S, Zattara M (1981) A sequence of postural movements precedes voluntary movement. Neurosci Lett 22: 263–270.

Bouisset S, Zattara M (1983) Anticipatory postural movements related to a voluntary movement. In Space Physiology, Cepadues Pubs, 137–141.

Carlsöö S (1962) A method for studying walking on different surfaces. Ergonomics, 5: 271–274.

Cordo PJ, Nashner LM (1982) Properties of postural movements related to a voluntary movement. J Neurophysiol 47: 287–303.

Corcos DM, Agarwal GC, Flaherty BP, Gottlieb GL (1990) Organizing Principles for Single-Joint Movements. IV. Implications for Isometric Contractions. J Neurophysiol 64: 1033–1042.

De Koning JJ, Van Ingen Schenau GJ (2000) Performance-Determining factors in speed skating. In Biomechanics in sports, IX (Zatsiorsky VM, ed.). 232–246

Flanagan JR, Wing AM (1995) The stability of precision grip forces during cyclic arm movements with a hand-held load. Exp. Brain Res, 105: 455–464.

Friedli WG, Cohen L, Hallett M, Stanhope S, Simon SR (1988) Postural adjustments associated with rapid voluntary arm movements. II. Biomechanical analysis. J Neurol Neurosurg Psych 51: 232–243.

Gaudez C, Le Bozec S, Richardson J (2003) Environmental constraints on force production during maximal ramp efforts. Archiv Physiol Biochem 111: 41.

Gaughran GRL. Dempster WT (1956) Force analyses of horizontal two-handed pushes and pulls in the sagittal plane. Human Biol 28: 67–92.

Gelfand IM, Gurfinkel VS, Tsetlin ML, Shik ML (1966) Problems in analysis of movements. In Gelfand IM, Gurfinkel VS, Fomin SV and Tsetlin ML (Eds). Models of the structural functional organisation of certain biological systems (American translation, 1971) 330–345. MIT Press, Cambridge, Mass.

Ghez C, Gordon J (1987) Trajectory control in targeted force impulses. I. Role of opposing muscles. Exp Brain Res 67: 225–240.

Ghez C, Gordon J, Ghilardi MF (1995) Impairments of reaching movements in patients without proprioception. I. Spatial errors. J Neurophysiol 73 :347–60.

Gordon J, Ghez C (1987) Trajectory control in targeted force impulses. II. Pulse height control. Exp Brain Res 67:241–252.

Grieve DW (1979) Environmental constraints on the static exertion of force: PSD analysis in task design. Ergonomics 22: 1165–1175.

Horak FB, MacPherson JM (1995) Postural orientation and equilibrium. Handbook of Physiology. Oxford University Press, New York, 255–292.

Johansson RS, Westling G (1984) Roles of glabrous skin receptors and sensorimotor memory in automatic precision grip when lifting rougher or more slippery objects. Exp Brain Res 56: 550–564.

Johansson RS, Westling G (1987) Signals in tactile afferents from the fingers eliciting adaptive motor responses during precision grip. Exp Brain Res 66(1):141–54.

Kroemer KHB (1974) Horizontal push and pull forces exertable when standing in working positions on various surfaces. Applied Ergonomics 5: 94–102.

Latash ML (1993) Control of Human Movement. Human Kinetics, Champaign, Il 1–380.

Lacquaniti F. Maioli C (1994) Coordinate transformations in the control of cat posture. J Neurophysiol 72: 1496–1515.

Lacquaniti F. Maioli C (1994) Independent control of limb position and contact forces in cat posture. J Neurophysiol 72: 1476–1495.

Lanshammar H, Strandberg L (1981) The dynamics of slipping accidents. J Occup Accid 3: 153–162.

Le Bozec S, Bouisset S (2004) Does postural chain mobility influence muscular control in sitting ramp pushes? Exp Brain Res 158: 427–437.

Le Bozec S, Goutal L, Bouisset S (1996). Are dynamic adjustments associated with isometric ramp efforts ? In Gantchev G.N., Gurfinkel V.S., Stuart D., Wiesendanger M., Mori S. (Eds). Motor Control Symposium VIII, Bulgarian Academy of Sciences, 140–143.

Le Bozec S. Goutal L. Bouisset S (1997) Dynamic postural adjustments associated with the development of isometric forces in sitting subjects. CR Acad Sci Paris 320: 715–720.

Le Bozec S, Lesne J, Bouisset S (2001) A sequence of postural muscle excitations precedes and accompanies isometric ramp efforts performed while sitting. Neurosci Lett 303: 72–76.

Lee WA, Patton JL (1997) Learned changes in the complexity of movement organization during multijoint, standing pulls. Biol Cybern 77: 197–206.

Lino F, Duchêne JL, Bouisset S (1992) Effect of seat contact area on the velocity of a pointing task. In Bellotti P. Capozzo A (eds) Biomechanics, Universita La Sapienza, Rome, 232.

Lino F (1995) Analyse biomécanique des effets de modifications des conditions d'appui sur l'organisation d'une tâche de pointage exécutée en posture assise. Thèse de Doctorat d'Université, Orsay, 1–211.

Macpherson JM. (1988) Strategies that simplify the control of quadrupal stance. I Forces at the ground. J Neurophysiol 60: 204–217.

Macpherson JM (1988) Strategies that simplify the control of quadrupal stance. II Electromyographic activity. J Neurophysiol 60: 218–231.

Macpherson JM (1991) How flexible are synergies ? In Humphrey DR. Freund HJ (eds) Motor Control Concepts and Issues, 33–47.

Nigg B (1986) Biomechanics of running shoes. Human kinetics Pub. , Champaign, Il. 1–180.

Strandberg L (1983) On accident analysis and slip-resistance measurement. Ergonomics 26: 11–32.

Tisserand M (1985) Progress in the prevention of falls caused by slipping. Ergonomics 28: 1027–1042.

Vandervael F (1956) Analyse des mouvements du corps humain. Paris: Maloine. 1–155.

Westling G, Johansson RS (1984) Factors influencing the force control during precision grip. Exp Brain Res 53 (2): 277–284.

Whitney RJ (1957) The strength of the lifting action in man. Ergonomics 1: 101–128.

Wing AM (1996) Anticipatory control of grip force in rapid arm movement. In Wing AM, Haggard P, Flanagan JR (Ed). Academic Press, London. 301–324.

Wing AM, Haggard P, Flanagan JR (1996) Hand and brain. The neurophysiology and psychology of hand movements. Academic Press, London. 1–513.

II. CONTROL OF RHYTHMIC ACTION

4. TRAJECTORY FORMATION IN TIMED REPETITIVE MOVEMENTS

Ramesh Balasubramaniam

Sensorimotor Neuroscience Laboratory, School of Human Kinetics, University of Ottawa, Canada

Abstract

In skills as diverse as piano playing or swinging a racquet in tennis, movements comprise a pattern that involves going to and away from the target, anecdotally referred to as attack and release. Although all such voluntary actions involve timing, timed repetitive movements involve bringing an end effector periodically to a certain location in the workspace, in relation to a sensorimotor event. Research in this area has involved the characterization of synchronization errors, identification of sources of variability in synchronization, and determination of neural structures involved in organizing such behavior. While much is known about the timing errors made while synchronizing with respect to external beat, not much is understood about what kind of movement trajectories are needed for timing accuracy. In this chapter, I review some recent work that links the ideas from the trajectory formation literature to what we currently know about timing accuracy in repetitive movements. Additionally, I present a paradigm that offers to bring together the dynamical systems approach with the information processing accounts of movement timing.

Introduction

There are two well identified traditions in movement timing research: the information processing approach and the dynamical systems approach. In the former, time is considered to be mental abstraction that is represented independent of any particular effector system (Wing, 2002; Vorberg & Wing, 1996). In this view, our ability to carry out an action such as playing the piano or hitting a ball at various speeds, to speak or draw fast or slow, depends on central timing processes. Time is represented independent of the motor apparatus, although it is generally understood that central timing processes might indeed make contact with the motor system. Said differently, according to this approach the central timing processes are functionally contained in that they do not need any particular effector system to be instantiated. While the timing processes may be set to initiate movements at certain times these movements' other parameters, such as force, amplitude or direction, can be specified independently (Semjen, Schulze & Vorberg, 2000).

In the dynamical systems approach, timing is considered to be an emergent property of the organizational principles (i.e., dynamical equations of motion) that govern coordinated action (Turvey, 1990; Kelso, 1995; Yu & Sternad, 2003). Thus the characteristic timing of an action is part and parcel of other movement dimensions of that action, such as frequency or its dynamical equivalents, stiffness and damping. A rhythmic activity such as piano playing may be carried out with regular timing but that is a consequence of a dynamical regime specifying a sequence of finger movement directions and amplitudes under particular stiffness constraints. In this approach, time as such is not an explicitly controlled variable, but follows from dynamical equations of motion and their parameter settings (for review see Schöner, 2002). The CNS does not deal with the abstract notion of time without reference to the moving parts of the body. So, for example, the control of timing in the production of a musical pattern may thus be said to follow from the effector system used to implement movement and its interaction with the environment. The key concept in the dynamical systems approach is that like all physical systems, brains and indeed behavior are governed by the laws of motion and change. The idea here is that *atomisms* or sub-symbolic units of behavior, emerge into organized, stable and meaningful structures on the basis of simple extremum principles or lawful constraints (Turvey, 1990). For example, bringing an end effector to a specific point in the workspace in time involves the collective action of several neuromuscular events whose macroscopic stability leads to the production of stable timed rhythmic behavior.

While the two approaches have divergent views on the nature of how temporal information is made available to and treated by the nervous system, they have also employed very different experimental paradigms (Wing & Beek, 2002; Balasubramaniam, Wing & Daffertshofer, 2004). The information processing approach has largely been concerned with the study of the synchronization event itself (Vorberg & Wing, 1996). Of special interest has been the statistical relationship between adjacent timing intervals in a sequence. The variability in the timing element of these movements has provided clues as to how the nervous system organizes movement onsets, arrivals or departures with respect to a specified meter(internal or external), with respect to successive arrivals, and in response to perturbations in phase and period (Repp, 2001).

On the contrary, the dynamical systems approach has looked at movement trajectories and their stability with respect to keeping with an external beat. Interesting experimental paradigms have looked at the stability of the phase of the movement of the end effector (1) with respect to an external event (Kelso, Delcolle & Schöner, 1990) or (2) with respect to another limb (Haken, Kelso & Bunz, 1985; Swinnen, 2002). Unearthing the nature of the "attractor" that keeps the movement of the effector in a phase relationship with the event in question has been a key question driving this approach. For example, what kind of attractor mechanism governs the behavior when one flexes the index finger to synchronize with a beat or syncopate to a beat (flexes the finger midway between two beats)? In the theory of dynamical systems, two well known attractors have been commonly used to describe timed repetitive movements: (1) The point attractor or stable fixed point (nearby trajectories converge onto a point; e.g., the equilibrium point of a mass-spring system) and (2) the periodic attractor or a limit cycle (trajectories converge onto a closed orbit; e.g., the periodic oscillations of a pendulum with an escapement to sustain the oscillations). The stability and variability of these movement patterns (along with their proclivities towards stable states) have provided clues as to the preferences shown by the nervous system in organizing rhythmic behavior. Such models have shown considerable success in explaining the relative stability of one regime with respect to another (Riley & Turvey, 2002).

PACED RESPONDING AND INTERVAL PRODUCTION

Behavioral studies of motor timing commonly focus on relatively short intervals up to a few seconds to span the timescale of voluntary movements. As mentioned in the earlier section, a frequently used paradigm in the information processing approach involves repetitive responding to produce a series of inter-response intervals I_j. Experimental control over mean (I) is obtained by including a period at the beginning of a trial in which the subject synchronizes responses with an auditory pulse train with inter-pulse interval set to T. When the pulses stop the subject is instructed to continue at the same rate for a further 30 to 50 responses. During the unpaced phase it is found that subjects maintain mean (I) within a few milliseconds of T, but with variability that increases with mean (I), a phenomenon first reported by Stevens (1886). A key characteristic of unpaced responding (also called continuation) is that successive I_j, I_{j+1} are negatively correlated between zero and minus one half. A theoretical account for this phenomenon was proposed by Wing and Kristofferson (1973). They suggested a hierarchical two-level model in which intervals generated by a central timer C_j are subject to delays in motor implementation D_j before the occurrence of observable responses. This model is henceforth referred to as the Wing & Kristofferson (W-K) model. The details of the working of the W-K model are presented in Figure 1.

LIMIT CYCLE OSCILLATORS AND THE W-K MODEL

Following a suggestion by Schöner (1994), it has been assumed that the lag-one autocorrelation effect predicted by the W-K model can be accounted for by the modulation of stiffness and damping of an autonomous limit cycle oscillator. A formal attempt has been made to account for empirically observed patterns of temporal variability in the W-K model with autonomous limit cycle oscillators. Daffertshofer (1998)—following an earlier suggestion of Schöner (1994)—examined, both analytically and numerically the minimal conditions under which limit cycle models with noise consistently produce a negative lag-1 serial correlation between consecutive periods of oscillation (with a value between 0 and -0.5). Contrary to earlier intuitions, he showed that a single (autonomous) limit cycle oscillator that is

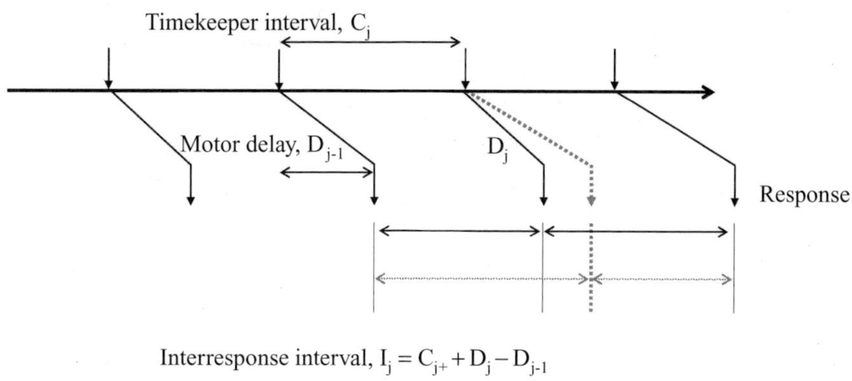

FIGURE 1. The operation of the Wing-Kristofferson (W-K) timing model. Variable timekeeper intervals subject to random motor implementation delays result in inter-response intervals that are negatively autocorrelated at lag 1 and bounded between zero and negative one half.

stochastically forced by (additive or multiplicative) white or by colored noise cannot produce the desired period correlation but results in phase diffusion, except under conditions of unrealistic stiffness values. In order to obtain reliable negative correlations, it is necessary either to introduce two conveniently placed interacting noise sources (regressing to the original WK model), or to add a second oscillator that is coupled to the limit cycle oscillator of interest (as a forcing function), thus stabilizing its phase. Thus if one were to take an oscillator based approach to account for the W-K results, a different strategy is needed. One obvious way to bring the two paradigms together is to look at movement trajectories of timed movement (Balasubramaniam et al, 2004; Delignieres et al, 2004).

Despite ideological differences between the two approaches, it is generally understood that control of timed repetitive actions should satisfy two goals: one directed at phase (precision and accuracy in timing) and the other at period (organization of movement parameters to meet interval requirements). What might be the constraints that would drive the requirements of a model that combines the two approaches? And more importantly, what kind of movement trajectories do we need to produce accurate movement timing?

Timing research has also historically paid little attention to the literature on trajectory formation. This is partly because models of trajectory formation and optimization have looked largely at discrete movements such as aiming and pointing. In discrete aiming movements, an important principle control principle is that of smoothness, based on jerk or the third derivative of position (Flash & Hogan, 1985). A sinusoidal trajectory (symmetric in position and velocity in the out and back phases) is a maximally smooth movement in that it minimizes the mean squared value of jerk (Flash & Hogan, 1985; Wann et al, 1988). It has been shown that the movement trajectories that have different velocity profiles in the two phases of a movement (hence asymmetric) typically have higher values of mean squared jerk (Nagasaki, 1991).

CEREBELLUM AND TIMING

A recent finding about the cerebellum's role in event timing and repetitive response production (Spencer, Zelaznik, Diedrichsen & Ivry, 2003) offers an interesting perspective on the two approaches. Spencer and colleagues found that patients with cerebellar damage (uni- and bilateral) could perform implicit timing tasks such as air tapping or circle drawing with little difficulty. However their ability to perform tasks with clearly defined temporal landmarks such as tapping to a beat with surface contact was quite seriously compromised. They suggest that the cerebellum, which is considered essential in setting and representing explicit temporal goals, plays a less important role in continuous movements. They argue that timing in continuous tasks is an emergent property that arises from the interactions of the neuromuscular system with the environment, without explicit temporal representations that involve the cerebellum. Spencer et al (2003) also suggested that "timing" in continuous movements (in the absence of cerebellar involvement) is likely to originate from a trajectory optimality criterion such as minimization of jerk.

MODES OF COORDINATION

There are two basic modes of coordinating movement with respect to an external metronomic event. They are (1) *Synchronization*: e.g. flexing the finger to strike on the beat and (2) *Syncopation*: e.g. flexing to strike off the beat or midway between beats, commonly seen in jazz. In musical contexts, syncopation is the more difficult skill and at higher frequencies shows an involuntary switch to synchronization. The skill is sometimes trained by redefining the focus of the task as extending the finger on the beat (Kelso et al, 1990). Thus flexion off the beat is achieved as a secondary consequence. In several laboratory studies it has been shown that extending on the beat is more stable than flexing off the beat, especially at higher frequencies, though not as stable as flexing on the beat (Kelso et al, 1998; Carson & Riek, 1998). Hence the definition of coordination with respect to an external metronome (Aschersleben & Prinz, 1995; Vorberg & Wing, 1996) should include not only task goals (synchronize vs. syncopate) but also motor goals (flexion vs. extension or pronation vs. supination).

Repeated to and fro movement is often approximately sinusoidal in form and hence assumed to be symmetric in the sense that the form and velocity of movement is similar in the out and back phases. This suggests constancy or symmetry of movement kinematics in the two phases. Symmetry in form is found even though the muscle activation required in each phase may be quite different due to dynamic factors such as the effects of gravity (Vallbo & Wessberg, 1993), unequal muscle forces (Cheney, Fetz & Mewes, 1991) and different sensori-motor cortical activation patterns (Yue et al, 1998). This symmetrical movement form has been used in several modelling efforts that have attempted to capture an oscillator description of finger movements, often involving limit cycles (Kay et al, 1991).

EVIDENCE AGAINST SYMMETRICAL TRAJECTORIES

In an experiment involving synchronization and syncopation to an external auditory metronome, Balasubramaniam et al (2004) have shown that the nervous system produces movement trajectories that are asymmetric with respect to time and velocity in the out and return phases of the repeating movement cycle (as shown in Figure 2). This asymmetry is task specific and is independent of motor implementation details (flexion vs. extension). Unpaced trajectories, however, do not show this asysmmetry in movement times or velocity in either direction of motion. Additionally, they found that timed trajectories are less smooth (higher mean squared jerk) than unpaced ones.

The mean squared jerk for the movements with a metronome present was much higher than the unpaced ones (as shown in Figure 3). Thus the timed movements showed a shorter, faster movement towards the target followed by a slower and longer movement away from the target. For example, in the condition flexing on the beat (fON), the flexion phase was shorter and faster than the extension phase, but the converse was true in the extend on the beat condition (eON). The trajectory in the Flexing off the beat (fOFF) condition resembled fON and not eON (which it is believed to be functionally equivalent to).

Additionally, negative correlations that were greater than -0.5 were observed between synchronization timing error and the movement time of the ensuing return phase, suggesting that late arrival of the finger is compensated by a shorter return phase and conversely for early arrival. Balasubramaniam and colleagues suggest that movement asymmetry in repetitive timing tasks helps satisfy requirements of precision and accuracy relative to a target event. Trajectory asymmetry was present in all conditions where subjects had to synchronize to an auditory metronome. In all the metronome paced conditions, subjects made more rapid movements of shorter duration towards the target and slow movements in the return phase. The degree of this asymmetry and consequently, mean squared jerk, decreased at higher metronome frequencies. In general, greater trajectory asymmetry was associated with better timing accuracy. Additionally, relative asynchrony (early or late arrival) was negatively correlated with the following slow phase.

It is interesting to note that duration of the "to" phase (such as flexion in fON), varies much less than the duration of the "away" phase across frequency conditions. One might suppose that the relative invariance of the "to" phase duration might underlie the changes in durational asymmetry of the movement trajectories. But a careful look at the correlations implicates the existence of a closed-loop control mechanism. Open-loop models of timing such as the Wing-Kristofferson (W-K)model (Wing & Kristofferson, 1973) predict that, in the absence of an external metronome, successive intervals between responses tend to exhibit a long and short alternation, resulting in a negative correlation that is theoretically bounded by zero and negative one half. The existence of a correlation between cycles greater than -0.5, as reported by Balasubramaniam and colleagues, suggests the presence of error correction or closed-loop control

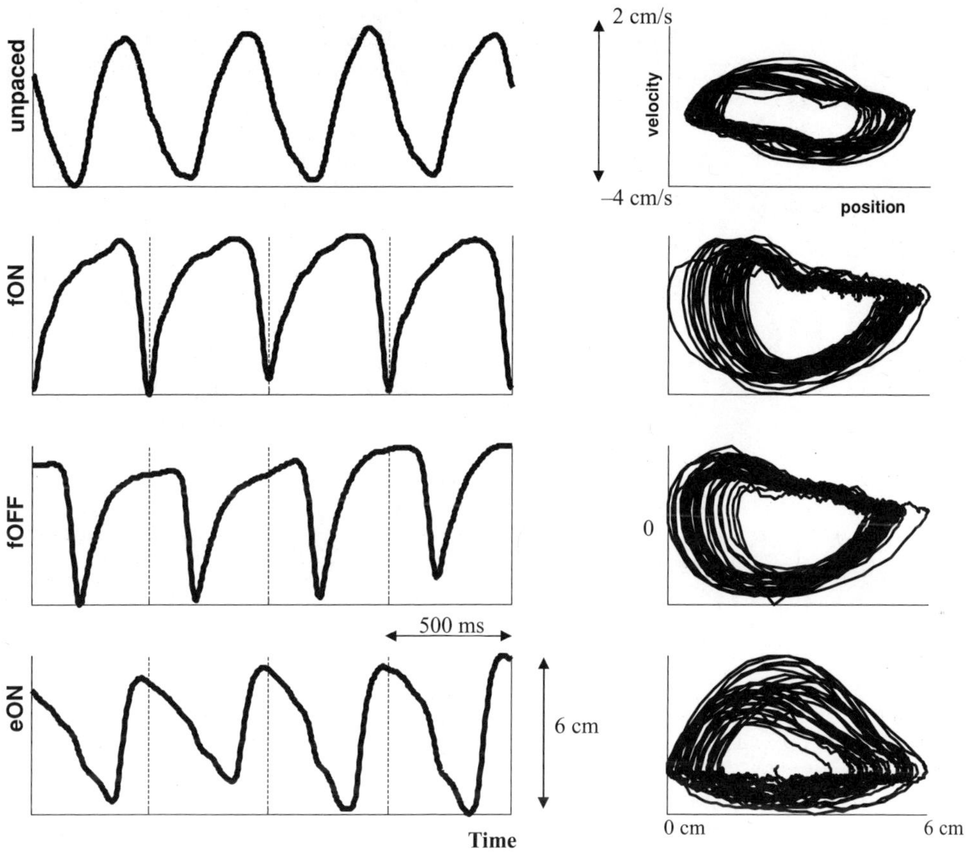

FIGURE 2. Asymmetry in movement trajectories. The left hand panel shows four cycles of displacement from a sample trial of a subject in the unpaced condition followed by fON, fOFF and eON. The dotted lines indicate the metronome event. The right hand panel shows the corresponding phase plots (position × velocity). Notice that the kinematic traces are symmetrical about flexion and extension in the unpaced condition and not so in the others. Also note that while fON and fOFF have similar extension/flexion profiles, eON is different.

(Vorberg & Wing, 1996; Pressing, 1999), which is characteristic of phase locking. It is important to mention here that the correlation that I have described is different from that used in the W-K model. While Balasubramaniam et al (2004) showed possible correctional mechanisms between relative asynchronies and the following movement phase, the W-K model refers to correlations between successive intervals. I am suggesting here that the trajectory asymmetry described here might provide a basis for and facilitate error correction.

It has been demonstrated that neural activity in proprioceptive pathways is scaled with the velocity of the movement (Delong et al, 1985; Gandevia & Burke, 1992; Grill & Hallett, 1995 & Matthews, 1991). Thus the modulation of velocity in timed movements (Turner et al, 1998) might be an active strategy employed by the CNS to detect proprioceptive sensory information. I would like to argue that high velocity movements towards the target may provide perceptual information relevant to phasing (accuracy in synchronization) and the slower return phase accommodates error correction and period adjustment. The general reduction of timing errors for higher movement frequencies or shorter time intervals (Aschersleben & Prinz, 1995), might also be related to movement velocity. Further experimentation in this area is required to clarify the role of movement velocity in the proprioceptive regulation of timing (Drewing et al, 2004).

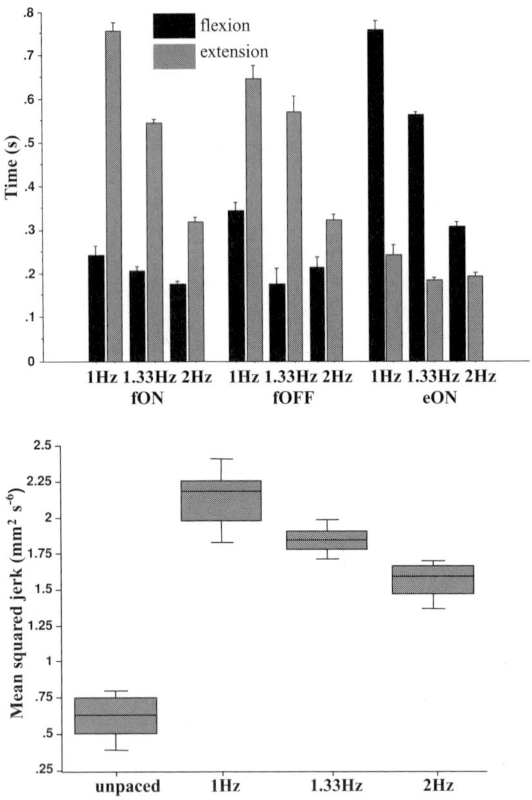

FIGURE 3. Statistical tests of asymmetry. The upper panel shows mean flexion and extension times for the fON, fOFF and eON conditions are plotted for each frequency. The lower panel shows that the mean squared jerk was significantly higher for the timed repetitive movement trajectories than the unpaced ones, with the slowest of the timed movements (most asymmetric) exhibiting the highest jerk.

To return to the argument made earlier, these results also suggest limitations on autonomous limit cycle oscillators as models of timed repetitive movements, because they are inherently symmetric. Interestingly, such limit cycle models have not been able to account for a fundamental aspect of timed movements that is the correlational structure between cycles as predicted by the W-K model (Daffertshofer, 1998). An oscillator model of timed repetitive movements (e.g., Beek et al, 2002) will have to take into account both the movement asymmetry and the correlational structure. It would be interesting and useful to see the development of models sensitive to the differing properties of each phase of the movement that also consider the optimization criteria for flexion and extension separately.

A starting point for such work might be to look at parameters like jerk, in addition to stiffness and damping separately for flexion and extension. Another avenue for further research might be to look at the optimization with respect to signal dependent noise present when issuing motor commands to move the finger (or an end effector) back and forth.

As noted earlier, the trajectory in the fOFF condition was more similar in form to fON than to the eON condition. It has been assumed following the experiments of Kelso et al (1995) that eON could actually be an alternative strategy for syncopation by fOFF. These results suggest that the functional similarities and differences between eON and fOFF at both the behavioral (Carson et al 1998) and cortical levels (Kelso et al, 1995) require a closer look.

FUNCTIONAL AND NEURAL IMPLICATIONS

It has been suggested by Spencer et al (2003) that the cerebellum, which is considered essential in setting and representing explicit timing goals, plays a less important role in continuous movements such as those presented here. Hence, it has been argued that timing in continuous tasks is an emergent property that arises from the interactions of the neuromuscular system with the environment, without explicit temporal representations that involve the cerebellum. Systemic modulations of parameters such as stiffness and damping that are not mapped directly onto specific neural or anatomical structures are implicated in the production of regular timed sequences of action (for review see Beek et al, 2002). Spencer et al (2003) also suggested that "timing" in continuous movements (in the absence of cerebellar involvement) is likely to originate from an optimality criterion such minimization of jerk. Here it is shown that jerk minimization which works well in the case of discrete movements such as spatial aiming might not be important in the control of timing in rhythmically paced movements. It is postulated that the alternating directions of movement with high and low velocity phases provide contrast in acceleration patterns that are useful landmarks for sensory (proprioceptive) regulation of timing. Thus the informational basis for timed action arises from the action itself. Further studies of cerebellar patients should address the issue of perceptual regulation of timing more carefully (Bracewell, Balasubramaniam & Wing, 2005).

Conclusions

The body of work that has been reviewed and presented here shows the benefits of combining two contrasting approaches (Wing & Beek, 2002) to timing:

discrete *event based* approaches that have looked at errors and their correction in synchronization, and *continuous* approaches that have almost exclusively dealt with the stability of movement trajectories. The question of what kind of optimality principles (Harris & Wolpert, 1998) are used by the CNS during trajectory formation in timed repetitive movements that satisfy constraints of accuracy and period stability is likely to be an important avenue for future research. Interesting future experimental methods would include studying trajectory formation in timed repetitive movements in the context of perturbations involving elastic and viscous force fields. This would reveal the relative importance of position and velocity based information in the regulation of timing. The question of what kind of oscillator (forced or unforced) model would account for W-K results still remains.

Acknowledgements

This work was supported in part by the Medical Research Council, UK and by an Initiation of New Research Direction (IRND) grant awarded by the University of Ottawa. I wish to thank Alan Wing, Andreas Daffertshofer and Andras Semjen for valuable discussions.

References

Aschersleben G & Prinz W (1995). Synchronising actions with events: the role of sensory information. *Percept Psychophys* 57: 305–317.

Balasubramaniam R, Wing AM & Daffertshofer A (2004) Keeping with the beat: Movement trajectories contribute to movement timing. *Exp Brain Res* 159: 129–134.

Beek, PJ, Peper CE & Daffertshofer A. (2002) Modeling rhythmic interlimb coordination: beyond the Haken-Kelso-Bunz model. *Brain Cog* 48: 149–165.

Bracewell RM, Balasubramaniam R & Wing AM (2005) Interlimb coordination deficits in a case of cerebellar hemiataxia. *Neurology* 64: 751–752.

Carson RG & Riek S (1998). The influence of joint position on the dynamics of perception-action coupling. *Exp Brain Res* 121: 103–114.

Cheney PD, Fetz CE & Mewes K (1993) Neural mechanisms underlying corticospinal and rubrospinal control of limb movements. *Prog Brain Res* 87: 213–252.

Daffertshofer A (1998) Effects of noise on the phase dynamics of nonlinear oscillators. *Phys Rev E* 58: 327–338.

Delignieres D, Lemoine L & Torre K (2004) Time interval production in tapping and oscillatory motion. *Hum Mov Sci* 23: 87–103.

DeLong MR Crutcher MD & Georgopoulos AP (1985) Primate globus pallidus and subthalamic nucleus: functional organization. *J Neurophysiol* 53: 530–543.

Drewing K, Stenneken P, Cole J, Prinz W & Aschersleben G (2004). Timing of bimanual movements and deafferentation: implications for the role of sensory movement effects. *Exp Brain Res* 158: 50–57.

Flash T & Hogan N (1985) The coordination of arm movements: an experimentally confirmed mathematical model. *J Neurosci* 5: 1688–1703.

Harris, CM & Wolpert DM (1998) Signal-dependent noise determines motor planning. *Nature* 394: 780–784.

Gandevia SC & Burke D (1992) Does the nervous system depend on kinesthetic information to control natural limb movements? *Behav. Brain Sci.* 15: 614–632.

Grill S & Hallett M (1995) Velocity sensitivity of human muscle spindle afferents and slowly adapting type II cutaneous mechanoreceptors. *J Physiol* 489: 593–602.

Kay BA, Saltzman EL, Kelso JAS (1991). Steady-state and perturbed rhythmical movements: A dynamical analysis. *J Exp Psychol : Hum Percept Perform.* 17: 183–197.

Kelso JAS, Fuchs A, Holroyd T, Lancaster R, Cheyne D & Weinberg H (1998). Dynamic cortical activity in the human brain reveals motor equivalence. *Nature.* 392: 814–818.

Kelso, JAS., DelColle J & Schöner, G. (1990). Action-Perception as a pattern formation process. In M. Jeanerod (Ed.), Attention and Performance XIII, Hillsdale, NJ: Erlbaum, 139–169.

Kelso JAS (1995) Dynamic Patterns. Cambridge: MIT press.

Matthews PBC (1981) Muscle spindles: their messages and their fusimotor supply. In: *Handbook of Physiology. The Nervous System. Motor Control.* Bethesda, MD.

Nagasaki H (1991) Asymmetric velocity and acceleration profiles of human arm movements. *Exp Brain Res* 87: 653–661.

Pressing J (1999) The referential dynamics of cognition and action. *Psych Rev* 106: 714–747.

Repp BH (2001) Processes underlying adaptations to tempo changes in sensorimotor synchronization. *Hum Mov Sci* 20: 277–312

Riley MA & Turvey MT (2002) Variability and determinism in motor behavior. J Mot Behav 34: 99–125.

Schöner G (2002) Timing, clocks and dynamical systems. *Brain Cog* 46: 31–51.

Stevens LT (1886) On the time sense. *Mind* 11: 393–404.

Spencer RM, Zelaznik HN, Diedrichsen J & Ivry RB (2003). Disrupted timing of discontinuous but not continuous movements by cerebellar lesions. *Science* 300: 1437–1439.

Swinnen SP (2002) Intermanual coordination: from behavioral principles to neural-network interactions. *Nat Rev Neurosci.* 3: 350–361.

Turner RS, Grafton ST, Votaw JR, Delong MR & Hoffman JM (1998) Motor subcircuits mediating the control of movement velocity: a PET study. *J Neurophysiol* 80: 2162–2176.

Turvey MT (1990) Coordination. *Am Psychol* 48: 938–153.

Valbo AB & Wessberg J (1993) Organization of motor output in slow finger movements in man. *J Physiol (Lond)* 469: 673–691.

Vorberg D & Wing AM (1996) Modeling variability and dependence in timing. In Heuer H & Keele SW (eds). *Handbook of Perception & Action.* 181–262. Academic Press, San Diego.

Wann JP, Nimmo-Smith I & Wing AM (1988) Relation between velocity and curvature in movement: equivalence and divergence between a power law and a minimum-jerk model. *J Exp Psychol : Hum Percept Perform* 14: 622–637.

Wing AM (2002). Voluntary timing and brain function. *Brain Cog* 48: 7–30.

Wing AM & Kristofferson AB (1973) Response delays and the timing of discrete motor responses. *Percept Psychophys* 14: 5–12.

Wing AM & Beek PJ (2002) Movement timing—a tutorial. In Prinz, W and Hommel, B (Eds) *Attention and Performance XIX.* Oxford University Press. pp. 202–226.

Yue GH, Liu JZ, Siemionow V (2000) Brain activation during finger flexion and extension movements. *Brain Res* 856: 291–300.

Zelaznik HN, Spencer RM & Ivry RB (2002). Dissociation of explicit and implicit timing in repetitive drawing and tapping movements. *J Exp Psychol : Hum Percept Perform* 28: 575–588.

5. STABILITY AND VARIABILITY IN SKILLED RHYTHMIC ACTION—A DYNAMICAL ANALYSIS OF RHYTHMIC BALL BOUNCING

Dagmar Sternad

Department of Kinesiology, The Pennsylvania State University

Abstract

The task of rhythmically bouncing a ball in the air serves as a model system that addresses many fundamental questions of coordination and perceptual control of actions. The task is simplified such that ball and racket movements are constrained to the vertical direction and the ball cannot be lost. As such, a discrete nonlinear model for the kinematics of periodic racket motions and ballistic ball flight between ball-racket contacts was formulated which permitted a set of analyses and predictions. Most centrally, linear stability analysis predicts that the racket trajectory should be decelerating prior to ball contact in order to guarantee dynamically stable performance. Such solutions imply that small perturbations need not be explicitly corrected for and therefore provide a computationally efficient solution. Four quantitative predictions were derived from a deterministic and a stochastic version of the model and were experimentally tested. Results support that human actors sense and make use of the stability properties of task. However, when single larger perturbations arise, human actors are able to adjust their racket trajectory to correct for errors and maintain a stable bouncing pattern.

Bouncing a ball rhythmically with a hand-held racket up in the air is a good exemplary task to address many fundamental issues on the control of action: How are the hand's and racket's movements controlled such that the ball is contacted at the right time with the right velocity in order to achieve a desired ball amplitude? To control the hand's, or racket's movements the actor requires perceptual information about the ball trajectory as well as about his/her own hand and arm trajectory. Based on this information the racket contacts the ball, which will, in turn, determine the next ball trajectory, providing new information for the subsequent ball-racket contact. In this way the task forms a perception-action loop—the action entails what is perceived, and what is perceived entails the next action. Further, the movements of racket and ball are rhythmic and, provided perfect performance, are invariant across cycles. As such, the task performance can be regarded as stable and at a dynamical equilibrium. This perspective will be made explicit by a model of the ball bouncing task that provides a quantitative description of stable performance at equilibrium. We will pursue the hypothesis that skilled performers seek and exploit the stability properties of the task.

A number of independent research lines have already examined variants of ball bouncing or juggling. For instance, several studies in robotics adopted "paddle juggling" or ball bouncing as a test bed for developing control algorithms for an actuator manipulating an object in 1D, 2D, and 3D (Bühler & Koditschek, 1990; Bühler, Koditschek, & Kindlmann, 1994). Humans bouncing a suspended ball with a bat attached to a pendular manipulandum served as a window to examine the phasing between ball and bat as a function of extrinsic task constraints and intrinsic properties of

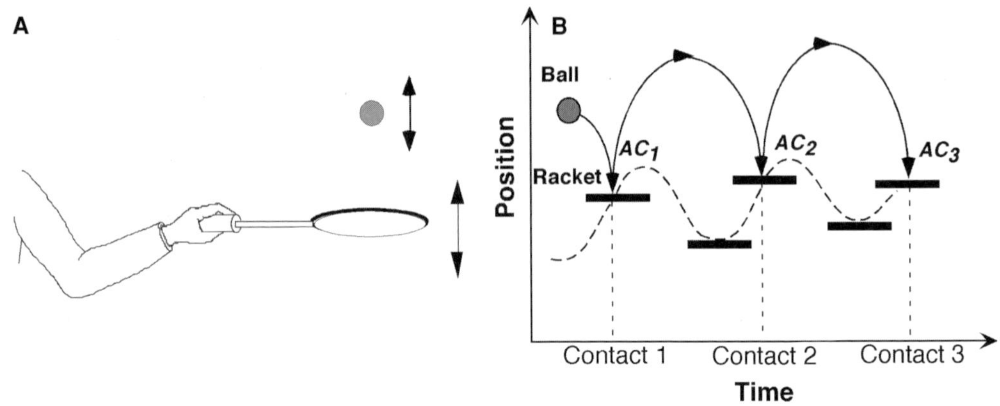

FIGURE 1. **A:** The model task. **B:** Exemplary time series of racket and ball trajectory and impact variable acceleration *AC* over three contacts.

ball and bat (Sim, Shaw, & Turvey, 1997). Bimanual juggling of three, five, or even seven balls was investigated in a series of studies by Beek and colleagues (Beek, 1989; Beek & van Santvoord, 1996). Inspired by Claude Shannon, who formulated a fundamental constraint on the timing of multi-ball-hand juggling (reported in Raibert (1986)), the studies further elaborated these constraints and examined how human jugglers find solutions within these constraints (Beek & Turvey, 1992). Yet, the most direct precursor for the present series of studies was work on the "hopping particle model" in the applied mathematics literature. This model consists of a periodically moving planar surface and a ball, both confined to the vertical direction, where the planar surface contacts the ball with inelastic impact and the ball obeys ballistic flight (Guckenheimer & Holmes, 1983; Tufillaro, Abbott, & Reilly, 1992). As such, the ball-racket system displays many basic features of nonlinear systems, such as stable states and a period doubling route to chaos. In analogy to this model, we conceived the experimental movement task such that the planar surface corresponds to the actor's racket bouncing the ball in the vertical direction. The basic assumption behind this approach is that the human-environment system constitutes a nonlinear dynamical system and the control of actions can be understood in this framework. More explicitly, we hypothesize that actors are sensitive to the stable states of a task system and learn to exploit them with increasing skill level. By such paralleling of model and task, analyses results of the model about stable states and bifurcations provide hypotheses that can be tested in human behavior.

The movement task is illustrated in Figure 1A, and Figure 1B shows the pertaining time series of racket and ball displacements. To comply with the model's assumptions both racket and ball motions in the experimental task are also confined to the vertical dimension. In both model and task, the ball trajectories follow ballistic flight with the gravitational constant $g = 9.81$ m/s^2. At impact, the ball bounces up in the air with a coefficient of restitution α, expressing the energy lost at each (instantaneous) impact.

We hypothesize that actors contact the ball such that they obtain *passive stability*. This term defines a type of performance that is resistant to perturbations, without adjusting the racket on a bounce-to-bounce basis. In the case of disturbances of the ball trajectory, the actor does not need to actively adjust the racket movements because the deviations die out by themselves. In the model, it can be shown that the condition for passively stable stationary ball movements is that the racket movement should impact the ball at times of negative racket acceleration, i.e., the racket trajectory is decelerating in the upward direction before and when contacting the ball. A detailed derivation and analysis of the model can be found in Dijkstra, Katsumata, de Rugy, and Sternad (2004) and Sternad, Duarte, Katsumata, and Schaal (2001). Thus, for the system to be passively stable in the stationary state, the critical parameter is the racket acceleration at impact. We denote this parameter by *AC*.

These predictions about *AC* were obtained from a local linear stability analysis of the model that revealed that one asymptotically stable state exists if *AC* is between zero and a negative value determined by g and α:

$$-2g\frac{(1+\alpha^2)}{(1+\alpha)^2} < AC < 0 \qquad (1)$$

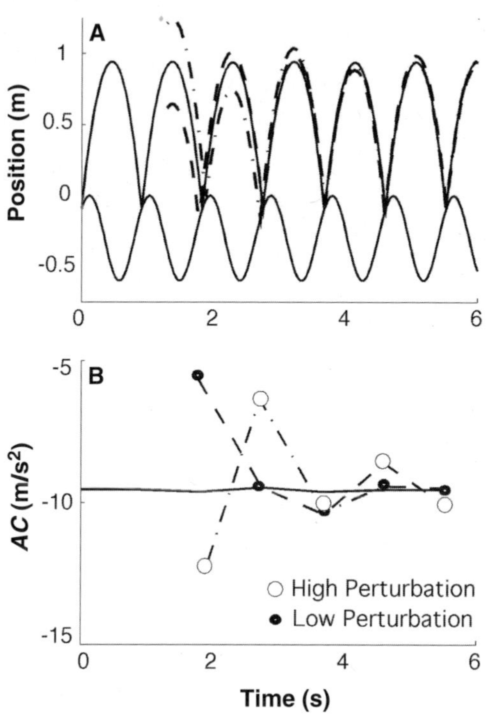

FIGURE 2. **A**: Three simulated time series of position of ball and racket trajectories with two perturbed trajectories shown by the dashed lines. **B**: Racket accelerations at contact *AC* over successive impacts for steady state (solid line) and perturbed impacts (dashed line).

When inserting the values for normal gravity ($g = 9.81$ m/s^2) and a typical coefficient of restitution ($\alpha = 0.50$), stability is obtained if *AC* is in the range between -10.90 m/s^2 and 0 m/s^2. Thus, once the actor has chosen the amplitude and frequency of the racket movement such that *AC* is in this range, the resulting performance is stable. Hence, we use the term "passive" stability.

To better appreciate this solution, Figure 2A shows a simulation where two ball trajectories are perturbed during the third cycle, one to a higher and one to a lower than previous amplitude. It can be seen that after a few cycles both ball trajectories converge back to the pre-perturbation amplitude. Importantly, the racket does not change its periodic trajectory. Figure 2B plots the corresponding racket acceleration values at contact, showing that the ball-racket contacts occur at a moment where the racket is decelerating in the upward direction, i.e., *AC* is negative in the stationary state as shown by the solid line (-9.5 m/s^2). The two dashed lines illustrate the *AC* values corresponding to the two perturbed trajectories. It can be seen that within three to four cycles the original *AC* value has been regained.

To emphasize that this passively stable regime is not trivial, two different strategies are conceivable. First, if actors maximize efficiency they should hit the ball at the moment of maximum upward velocity, as for a given racket amplitude the maximum ball amplitude is achieved. This moment of contact corresponds to zero *AC*. Second, the so-called mirror algorithm was developed and implemented on a robot system that juggled a ball in 2D and 3D (Bühler & Koditschek, 1990; Bühler et al., 1994; Sternad & Dijkstra, 2004). This strategy generates stable rhythmic bouncing movements, albeit with different means. This algorithm uses continuous visual information about the ball trajectory to generate racket movements that mirror the ball's movements. This control strategy leads to positive values of the impact parameter *AC*. In contrast, the passive stability strategy involves no such feedback but relies on the actor choosing a period and amplitude such that the *AC* is negative and within the range specified by equation (1). While all three strategies are feasible, we hypothesize that trained subjects will favor the passively stable regime. This solution necessitates less control and is computationally less expensive, leaving attentional resources free for other demands. In the following, a series of experiments is summarized that tested this and associated hypotheses that will be introduced below.

General Methods

Subjects performed the experimental task with a custom-made apparatus where they held a tennis racket in their dominant hand and bounced a ball up in the air. The ball was attached to a 1-m long boom rotating on a hinge joint to confine the ball's movements to a one-dimensional curvi-linear path (Sternad et al., 2001). Due to this fixture the ball could not be lost in the performance. Within the observed ball amplitudes the ball trajectory could be assumed to be linear, in close resemblance to the model's assumptions. An accelerometer attached to the rim of the racket directly measured the racket's acceleration. The coefficient of restitution α was experimentally determined and had the value $\alpha = 0.52$. Due to the attachment the ball's flight was no longer in normal gravitational conditions, and experimental determination of g yielded a value of 5.6 m/s^2. Hence, the range of stable solutions for these parameter values was between -6.16 and 0 m/s^2 (see eq 1). Subjects were instructed to bounce the ball rhythmically with a steady ball amplitude throughout the duration of a trial

(40 s) which typically comprised 50 to 70 ball-racket contacts.

The primary dependent variable was the acceleration of the racket at impact with the ball, *AC*. The average value across the approximately 50–70 contacts during one trial was used to compare subjects' performance with the model's predictions about passive stability. In addition, the amplitude of the ball trajectory was determined as the distance between the ball-racket impact and the subsequent peak of the ball trajectory. Variability of task performance was captured by the standard deviations of the amplitudes across one trial in *SDA*.

Predictions and Results

PREDICTION 1: PASSIVE STABILITY IN SKILLED PERFORMANCE

In Experiment 1 we tested how novice performers changed and improved their performance across a sequence of 40 trials, where each trial lasted 40 s (Dijkstra et al., 2004; Sternad & Dijkstra, 2004). This session provides approximately 30 minutes of practice, which was sufficient as the task was relatively easy for subjects. The hypothesis was that with increasing skill level, actors should attune to passive stability properties of the task and start to exploit them. The dependent measures were the average *AC* and the associated variability estimates *SDA* were calculated for each trial. Figure 3 illustrates with the data of one exemplary subject how *AC* and *SDA* changed across the experimental session. Variability decreased across the 40 trials indicating improvement of the task. Parallel to this improvement, the values of *AC* decreased from positive values towards negative values that asymptoted towards a value of approximately -3.5 m/s^2. This change in *AC* is consistent with Prediction 1: From initially unstable performance, the subject changed to racket-ball contacts that provide passive stability. Note that human performance with positive *AC* values is indeed possible and does not lead to a loss of pattern as would be predicted from purely passively stable performance. This means that human actors also have alternative strategies. It may be speculated that in the beginning of the practice subjects rely more strongly on continuous visual information about the ball and thereby use a mirror-like strategy that leads to a positive *AC* value at contact.

PREDICTION 2: DEGREE OF STABILITY

In addition to the local linear stability analysis, a non-local Lyapunov stability analysis permitted finer-grained predictions on the degree of stability for the given values of *AC* (Schaal, Sternad, & Atkeson, 1996; Sternad et al., 2001). If the calculated values for stability are compared with the observed variability, then a skewed U-shaped dependence of variability should be expected (see connected line in Figure 4). In Experiment 2 eight subjects, of different levels of practice, each performed five trials of steady-state bouncing with a self-chosen ball amplitude. The trial means of *AC* and standard deviations of ball amplitudes *SDA* served as the operationalized measures. Figure 4 plots

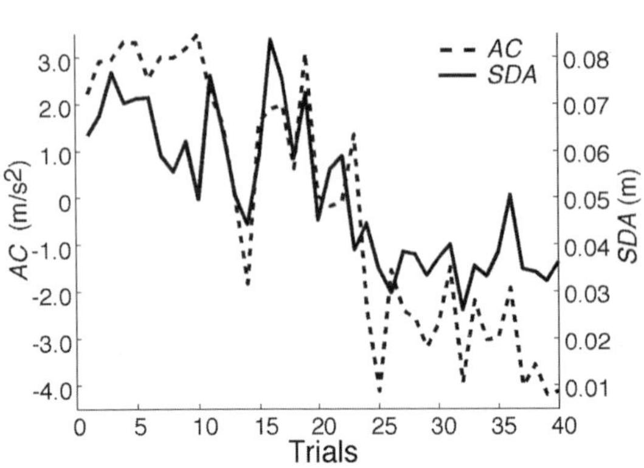

FIGURE 3. Mean acceleration at contact (*AC*) and variability of ball amplitude (*SDA*) of trials across a practice session of 40 trials.

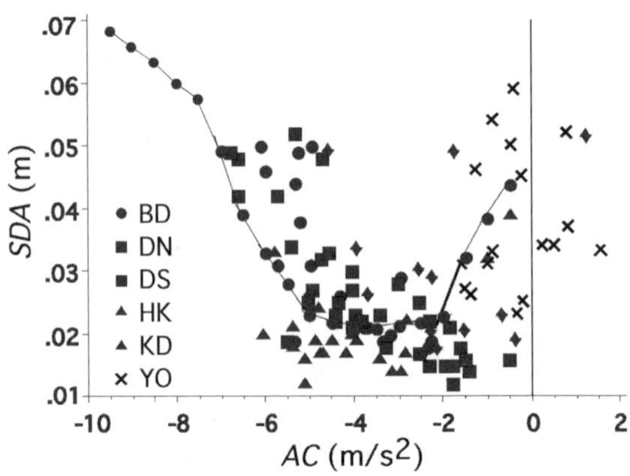

FIGURE 4. Variability, measured as standard deviations of ball amplitude (*SDA*), plotted against its corresponding average accelerations at contact (*AC*) for trials in 6 different subjects performing 15 trials each.

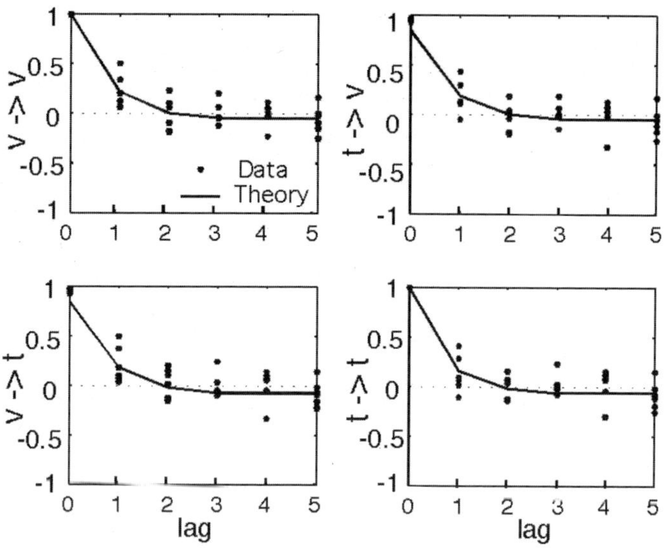

FIGURE 5. Auto- and cross-correlations between velocity of the ball (v) and time between contacts (t) for five lags.

SDA against *AC*. It can be seen that different subjects performed with different values of *AC*. Concomitantly and in line with the predictions, the magnitude of variability is a function of the average *AC* value. Further, it is interesting to note that subject HK was the most experienced performer with this apparatus, and he performed with *AC* values of approximately -4 m/s^2, where the Lyapunov analysis predicts lowest variability. In contrast, subject YO had no experience with racket sports at all. Several *AC* values are even positive, i.e., outside the stable range, and accompanied by relatively high fluctuations in the ball amplitude.

PREDICTION 3: COVARIANCE STRUCTURE DURING STEADY-STATE PERFORMANCE

In order to make finer-grained predictions about fluctuations during steady state performance and after small perturbations, the deterministic model was extended by adding a stochastic component (Dijkstra et al., 2004). Adding noise to the model introduces small perturbations to the state variables that die out due to the stability properties of the map, albeit with specific relaxation behavior. This behavior is captured in the correlation functions of the state variables of the ball bouncing model. When the same correlation functions are determined for the experimental data, model predictions for passively stable behavior can be tested. For this analysis the state variables were ball velocity after contact (v) and time between contacts (t). Their four auto- and cross-correlations are shown in Figure 5 for five lags (impacts). The predictions of the simulations are shown by the solid line, the data of five exemplary trials by solid dots. In general, the data conform with the predictions. The most important results are that (*i*) lag-1 for all four correlation functions are positive; (*ii*) all lags higher than 1 are zero. The fact that all correlation functions are similar is probably due it the fact that velocity noise dominates (for more detail see Dijkstra et al., 2004).

PREDICTION 4: RELAXATION OR CORRECTION AFTER LARGER PERTURBATIONS

Figure 2 illustrated how perturbations of the ball trajectories converge back to the stable regime without requiring racket adjustments when the model performs with passive stability. However, this behavior was demonstrated for small noise-like perturbations. How does the model and subjects' behavior compare when larger perturbations are applied and the individual returns to stable performance?

In Experiment 3 six subjects performed rhythmic ball bouncing with and without perturbations of the ball trajectory. Such experimental manipulations were only possible in a virtual set-up where the actor manipulated a real racket but the ball and its interactions with the racket only existed virtually. The racket movements were presented in real time on a 2D projection

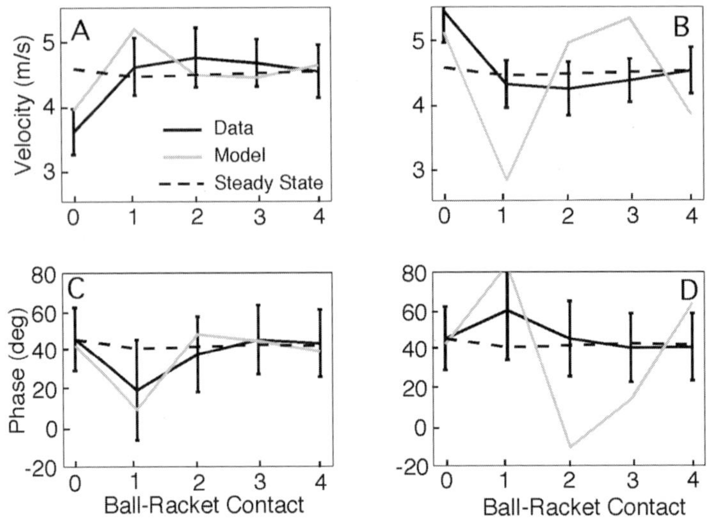

FIGURE 6. Response to perturbations in the model and the experimental data, measured in velocity of the ball at contact and the time between contacts across five contacts at and following the perturbation. **A**, **C**: Perturbations with α_P smaller than 0.50. **B**, **D**: Perturbations with α_P larger than 0.50.

screen where the simulated ball trajectories were also presented (for a detailed description see de Rugy, Wei, Müller and Sternad (2003)). In the unperturbed condition, α remained constant at 0.50 during the entire trial. In the perturbed trials, α was changed at every fifth impact to a random value α_P within the ranges 0.30 and 0.40, or 0.60 and 0.70, leaving it at 0.50 otherwise. One trial gave approximately 50 cycles and contacts and thus 10 perturbations for each trial. The experiment consisted of 15 trials per condition (40 s each trial).

The subjects' return patterns to stable bouncing after the perturbations are shown in Figure 6 by the solid black lines. Figures 6A and 6C show average values of velocity and phase for the perturbed ball-racket contact (0) and the subsequent contacts (1 to 4) when α_P values were smaller than 0.50. Figures 6B and 6D show the returns for α_P larger than 0.50. The dashed lines show subjects' values for unperturbed performance. The solid grey lines show the simulated relaxation patterns. While there is some qualitative congruence in Figure 6A and 6C, the subjects' data always showed a faster return to stable behavior. Especially for α_P values larger than 0.50, the model shows large oscillatory returns that have not reached equilibrium after 5 contacts. This pattern is not observed in the subjects who always return to equilibrium after approximately one or two contacts. This observation suggests an additional strategy that involves active corrective processes.

To elucidate how subjects reached equilibrium so fast, the continuous racket trajectories were parsed into individual cycles separated at the ball-racket impact. All cycles were sorted into C-1 to C-5, where C-1 is the first post-perturbation cycle, and then averaged by cycle number. Figure 7 summarizes the pattern of periods and amplitudes across post-perturbation cycles and perturbation condition. More precisely, it is the differences in period and amplitude at each post-perturbation impact compared to the average period and amplitude that is computed to facilitate a comparison across subjects who had different mean values of period and amplitude. Figure 7A shows that the racket periods were increased or decreased by approximately 0.15 s during C-1 in response to the perturbation for both smaller and larger α_P. In contrast, the amplitude modulations in Figure 7B showed no such difference across the cycles for the two α_P conditions. The small constant offset between the two α_P conditions is not statistically significant, as evident from the error bars.

These results unequivocally show that, when necessary, subjects modulate their racket trajectory to regain a regular bouncing pattern. Interestingly, the adjustments were made in the period only. De Rugy, Sternad, and colleagues (2003) presented a model in which these results were replicated with a discrete coupling

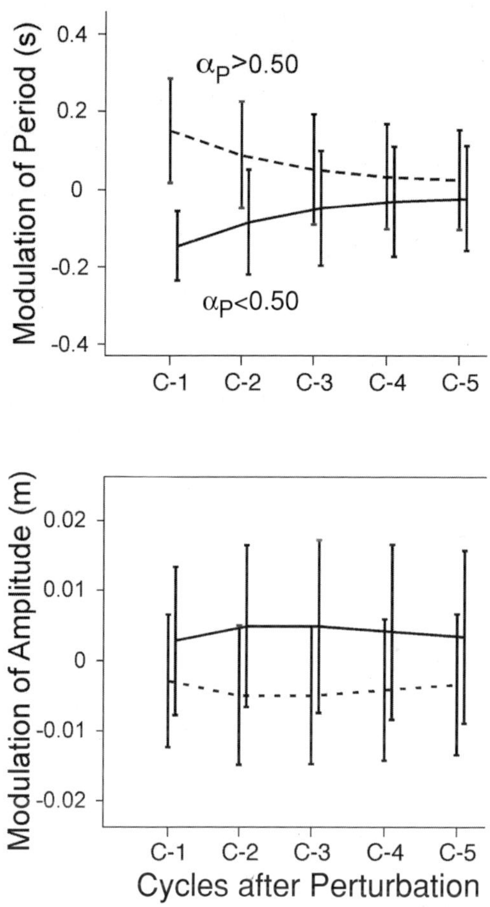

FIGURE 7. Period and amplitude modulations of the racket trajectory following perturbations. C-1 refers to the cycle directly after the perturbation. Solid lines connect data points for $\alpha_P > 0.50$, dashed lines for $\alpha_P < 0.50$.

between the racket movements and the ball trajectories at the moment of contact. Rhythmic racket trajectories were generated by an oscillator whose period was parameterized as a function of the velocity of the ball immediately following the contact. Adjustments were achieved by a resetting of the period directly following the impact as a function of the ball velocity after contact. The simulation replicated most of the features in the data.

In sum, this selective overview of experimental data and model analyses from a long series of studies showed that human actors are indeed sensitive to the stability properties of the task. During practice subjects learn to tune into these properties, probably to alleviate the computational demands on the control of actions. The variability in central task parameters such as ball amplitude parallels this attunement process and decreases with practice. Variability in expert performance also depends on the chosen values of racket acceleration, complying with more fine-grained predictions from a Lyapunov analysis. Further, fluctuations in contact variables in steady state performance show a covariance structure that qualitatively conforms with the one from a stochastic version of the model. However, when single perturbations are applied actors reveal an additional strategy that ensures a relatively fast equilibration to the stable bouncing pattern: Subjects modify the period of their racket movements almost immediately with the perturbed ball trajectory. This modulation follows visual information about the perturbed ball trajectory. This shows that despite the possibility to simply maintain an invariant racket trajectory once a stable pattern is established, actors continue to monitor the ball trajectory and flexibly modulate their actions as the task requires.

Acknowledgements

This work was supported by the grants from the National Science Foundation, PAC-0450218, and BCS-0096543, and the National Institutes of Health R01-HD045639. I would like to thank the many collaborators, but specifically Tjeerd Dijkstra, who have worked on this research and discussing issues over many years.

References

Beek, P. J. (1989). *Juggling dynamics.* Unpublished Doctoral Dissertation, Free University Press, Amsterdam.

Beek, P. J., & Turvey, M. T. (1992). Temporal patterning in cascade juggling. *Journal of Experimental Psychology: Human Perception and Performance, 18, 4,* 934–947.

Beek, P. J., & van Santvoord, A. A. M. (1996). Dexterity in cascade juggling. In M. L. Latash & M. T. Turvey (Eds.), *Dexterity and its development* (pp. 377–392). Mahwah, NJ: Erlbaum.

Bühler, M., & Koditschek, D. E. (1990). *From stable to chaotic juggling: Theory, simulation, and experiments.* Proceedings at the IEEE International Conference on Robotics and Automation, Cincinnati, OH, 1976–1981.

Bühler, M., Koditschek, D. E., & Kindlmann, P. J. (1994). Planning and control of robotic juggling and catching tasks. *International Journal of Robotics Research, 13,* 101–118.

de Rugy, A., Wei, K., Müller, H., & Sternad, D. (2003). Actively tracking "passive" stability. *Brain Research, 982, 1,* 64–78.

Dijkstra, T. M. H., Katsumata, H., de Rugy, A., & Sternad, D. (2004). The dialogue between data and model: Passive stability and relaxation behavior in a ball bouncing task. *Journal of Nonlinear Studies, 3*, 319–345.

Guckenheimer, J., & Holmes, P. (1983). *Nonlinear oscillations, dynamical systems, and bifurcations of vector fields.* New York: Springer.

Raibert, M. (1986). *Legged robots that balance.* Cambridge, MA: MIT Press.

Schaal, S., Sternad, D., & Atkeson, C. G. (1996). One-handed juggling: A dynamical approach to a rhythmic movement task. *Journal of Motor Behavior, 28, 2*, 165–183.

Sim, M., Shaw, R. E., & Turvey, M. T. (1997). Intrinsic and required dynamics of a simple bat-ball skill. *Journal of Experimental Psychology: Human Perception and Performance, 23, 1*, 101–115.

Sternad, D., & Dijkstra, T. M. H. (2004). Dynamical stability in the acquisition and performance of rhythmic ball manipulation: Theoretical insights with a clinical slant. *Journal of Clinical Neurophysiology, 3, 11*, 215–227.

Sternad, D., Duarte, M., Katsumata, H., & Schaal, S. (2001). Bouncing a ball: Tuning into dynamic stability. *Journal of Experimental Psychology: Human Perception and Performance, 27, 5*, 1163–1184.

Tufillaro, N. B., Abbott, T., & Reilly, J. (1992). *An experimental approach to nonlinear dynamics and chaos.* Redwood City, CA: Addison-Wesley.

6. THE DISTINCTIONS BETWEEN STATE, PARAMETER AND GRAPH DYNAMICS IN SENSORIMOTOR CONTROL AND COORDINATION

Elliot Saltzman

Department of Rehabilitation Science, Boston University, Boston MA, USA; Haskins Laboratories 300 George Street, New Haven, CT USA

Hosung Nam

Department of Linguistics, Yale University, New Haven, CT USA

Louis Goldstein

Haskins Laboratories, 300 George Street, New Haven, CT USA; Department of Linguistics, Yale University New Haven, CT USA

Dani Byrd

Department of Linguistics, University of Southern California, Los Angeles, CA USA

Abstract

The dynamical systems underlying the performance and learning of skilled behaviors can be analyzed in terms of state-, parameter-, and graph-dynamics. We review these concepts and then focus on the manner in which variation in dynamical graph structure can be used to explicate the temporal patterning of speech. Simulations are presented of speech gestural sequences using the task-dynamic model of speech production, and the importance of system graphs in shaping intergestural relative phasing patterns (both their mean values and their variability) within and between syllables is highlighted.

I. *Introduction*

What is being learned when we learn a skilled behavior? In our opinion, and in those of many other proponents of the dynamical systems approach to sensorimotor control and coordination, what is learned is the underlying dynamical system or coordinative structure that shapes functional, task-specific coordinated activity across actor and environment. But this begs the question of just what sort of a beast a dynamical system *is*. Fortunately, it is not too difficult to define one. Roughly, a dynamical system is a system of interacting variables or components whose individual behaviors and whose modes of interaction are shaped by laws or rules of motion. The focus of this chapter is on the types of variables that comprise a dynamical system, the types of laws or rules that govern changes of these variables over time, and the manner in which such changes can be related to processes of coordination and control in skilled behavior, with particular emphasis on temporal patterning in the production of speech.

State-, Parameter-, and Graph-Dynamics. Any dynamical system can be completely characterized according to three sets of variables—state-, parameter-, and graph-variables (Farmer, 1990; Saltzman & Munhall, 1992)—and the laws or rules that govern their respective dynamical changes over time. State-variables can be viewed as the system's active degrees of freedom, and are represented as the dependent or output variables of the set of autonomous differential or difference equations of motion that are used to describe the system. More specifically, a given n^{th} order dynamical system has n state variables and can be described, equivalently, by a single n^{th}-order equation of motion or by a set of n-1st order equations of motion, with one 1st-order equation of motion for each state variable. For example, 2nd order mechanical systems such as damped mass-spring systems or limit cycle pendulum clocks have two state variables, position and velocity; typical n^{th}-order computational (connectionist) neural networks have n state variables that are defined by the activation levels of each of the network's n processing nodes. State dynamics refers to the manner in which changes over time of the state variables are shaped by the "forces" (more technically, the state-velocity vector field) inherent to the system that are described by the system's equation(s) of motion. A system's parameters are typically defined by the coefficients or constant terms in the equation of motion. For example, these could be the mass, damping, and stiffness coefficients or the constant target parameter in a damped, mass-spring equation, the length and mass of a clock's pendulum, or the inter-node synaptic coupling strengths in a computational neural network. Parameter dynamics refers to the manner in which changes in parameter values are governed over time. In general, a system's parameters change more slowly than its state-variables, although this is not always the case. For example, a child's limb lengths and masses change at an ontogenetic timescale while children's skilled limb movements unfold in real-time. Similarly, a connectionist network's synaptic weights change over the timescale defined by the learning algorithm used to train the network to solve a given computational task; this learning timescale is typically much slower than the state-dynamic performance timescale of the activation state variables. It is possible, however, for system parameters to change on a timescale comparable to, or even faster than, the corresponding state-variables' timescale. For example, we can intentionally change the rate at which we reach toward a target, or even switch from one target to another, during the reaching motion itself.

The notion of a system's graph is a less familiar one, at least in the domain of sensorimotor control and coordination, than that of the system's set of state-variables or parameters. The graph of a system

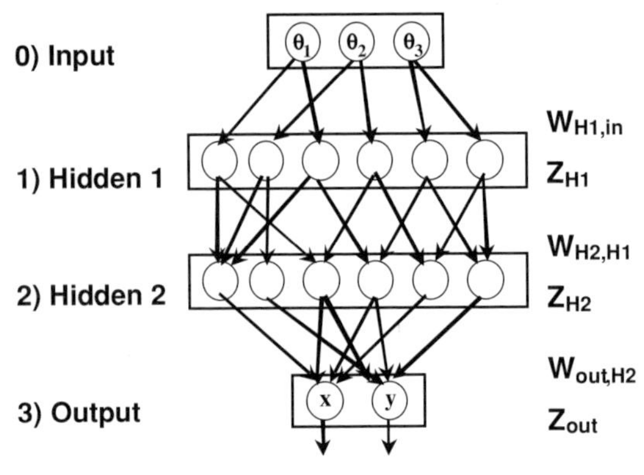

FIGURE 1. Graph of typical feedforward connectionist network. $W_{i,j}$: matrix of synaptic weights from j-layer cells to i-layer cells; Z_i: vector whose entries are activation output values of i-layer cells.

represents the "architecture" of the system's equation of motion, and denotes the parameterized set of relationships defined by the equation among the state-variables. For connectionist systems, the graph is simply the standard node+linkage diagram used to represent such systems (see Figure 1). For non-connectionist systems, the conceptual connection between a system's graph and its equation of motion is less straightforward, but becomes clearer when one realizes that a symbolically written differential or difference equation can be represented equivalently in pictorial form as a circuit diagram. The latter type of representation can be used to construct an electronic circuit to simulate the system on an analog computer, or to graphically define the equation in an application such at Matlab's Simulink for simulating the system on a digital machine. Figure 2 shows a circuit diagram used to simulate a 2nd-order damped mass-spring system using Simulink.

Graph dynamics refers to the manner in which the system graph changes over time. This can include changes in the systems's dimensionality, i.e., the number of active state variables, and in the structure of the state variable functions included in the system's equation of motion. In a behavioral context, the number of active state variables might change due to a decision or instruction to switch from unimanual to bimanual lifting of a given object, or due to the recruitment of the trunk in addition to the arm when reaching for a distant target. Relatedly, in connectionist systems trained by "constructivist" learning algorithms, nodes and linkages can be added or

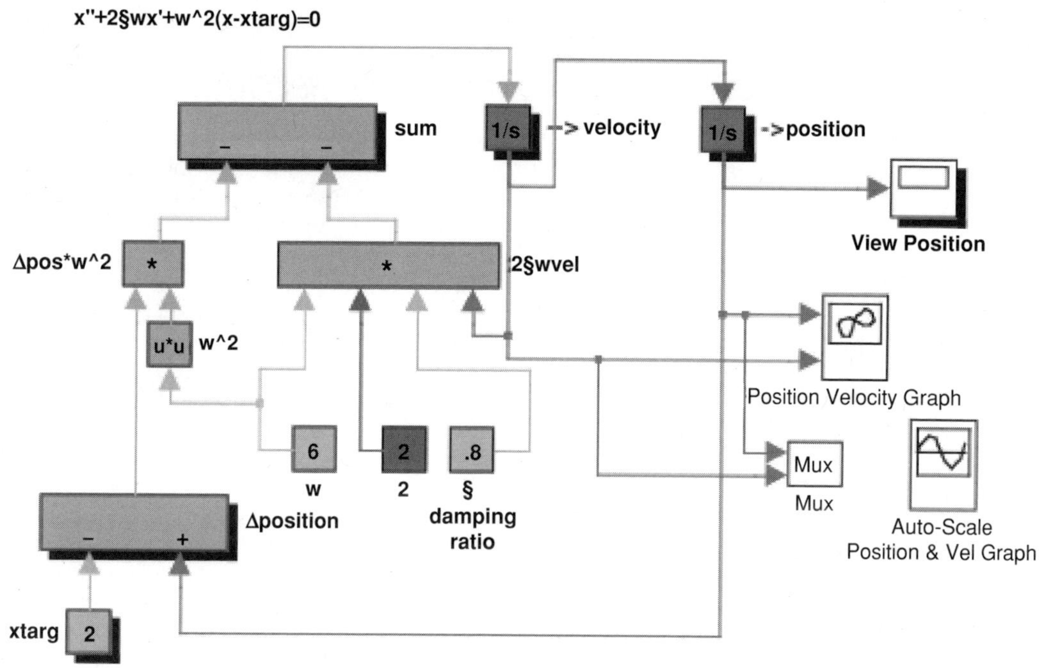

FIGURE 2. Simulink circuit diagram for 2nd-order damped mass-spring equation, with symbolic equation written in the upper left corner. x, x', and x'' denote position, velocity, and acceleration, respectively; §, w, and xtarg denote damping ratio, natural frequency, and target parameters, respectively; Δ pos = x–xtarg, and 1/s denotes the operation of integration.

deleted to create a network whose structural complexity is adequate for instantiating a "grammar" that is sufficient for learning particular classes of functions (e.g., Huang, Saratchandran, & Sundararajan, 2005; Quartz & Sejnowski, 1997). Additionally, the damping functions implemented during discrete point-attractor tasks such as reaching or pointing will be qualitatively different from those required to perform sustained rhythmic, limit cycle polishing or stirring tasks; similarly, interlimb coupling functions will be different for skilled performances of bimanual polyrhythms with correspondingly different *m:n* frequency ratios.

In the following sections, we will review some recent work of ours that highlights the role of system graphs in shaping the temporal patterning of speech. Our work is presented within the framework of the Task-Dynamic model of speech production (e.g., Saltzman 1986; Saltzman & Munhall 1989; Saltzman & Byrd, 2000), and we focus on the manner in which the relative timing of speech gestures, i.e., intergestural timing, within and between syllables (with respect to both mean intergestural time intervals and their variability) can be understood with reference to the system's underlying intergestural coupling structures.

II. *The Task-Dynamic Model of Speech Production: An Overview*

The temporal patterns of speech production can be described according to four types of timing properties: intragestural, transgestural, intergestural, and global. Intragestural timing refers to the temporal properties of a given gesture, e.g., the time from gestural onset to peak velocity or to target attainment; transgestural timing refers to modulations of the timing properties of all gestures active during a relatively localized portion of an utterance; intergestural timing refers to the relative phasing among gestures (e.g., between bilabial closing and laryngeal devoicing gestures for /p/, or between consonantal bilabial closing and vocalic tongue dorsum shaping gestures for /pa/); and global timing refers the temporal properties of an entire utterance, e.g., overall speaking rate or style.

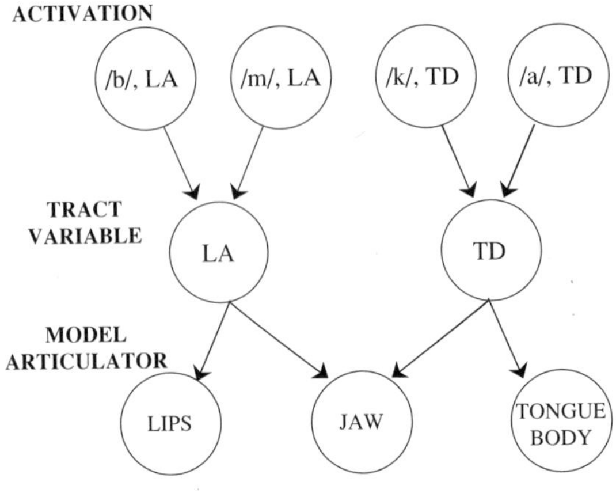

FIGURE 3. Three sets of state variables in the Task-Dynamic Model. Activation variable dynamics are defined at the intergestural level of the model; tract variable and model articulator dynamics are defined at the interarticulator level of the model. LA and TD denote lip aperture and tongue dorsum tract variables, respectively.

In the task-dynamic model of speech production, the spatiotemporal patterns of articulatory motion emerge as behaviors implicit in a dynamical system with two functionally distinct but interacting levels. As shown in Figure 3, the *interarticulator* level is defined according to both *model articulator* (e.g. lips & jaw) coordinates and *tract-variable* (e.g. lip aperture [LA] & protrusion [LP]) coordinates; the *intergestural* level is defined according to a set of *activation* coordinates. Invariant gestural units are posited in the form of context-independent sets of dynamical parameters (e.g. target, stiffness, and damping coefficients) and are associated with corresponding subsets of model articulator, tract-variable, and activation coordinates. Each unit's activation coordinate reflects the strength with which the associated gesture (e.g., bilabial closing) "attempts" to shape vocal tract movements at any given point in time. The tract-variable and model articulator coordinates of each unit specify, respectively, the particular vocal-tract constriction (e.g. bilabial) and articulatory synergy (e.g. lips and jaw) whose behaviors are affected directly by the associated unit's activation. The interarticulator level accounts for the coordination among articulators at a given point in time due to the currently active gesture set. The intergestural level governs the patterns of relative timing among the gestural units participating in an utterance and the temporal evolution of the activation trajectories of individual gestures in the utterance. The trajectory of each gesture's activation coordinate defines a forcing function specific to the gesture and acts to insert the gesture's parameter set into the interarticulatory dynamical system defined by the set of tract-variable and model articulator coordinates. Additionally the activation function gates the components of the forward kinematic model (from model articulators to tract variables) associated with the gesture into the overall forward and inverse kinematic computations (see Saltzman & Munhall, 1989, for further details).

In the original version of the model, the intergestural level used *gestural scores* (e.g. Browman & Goldstein, 1990) that explicitly specified the activation of gestural units over time and that unidirectionally drove articulatory motion at the interarticulator level. In these gestural scores, the shapes of activation waves were restricted to step functions for simplicity's sake, and the relative timing and durations of gestural activations were determined, until relatively recently, either with reference to the explicit rules of Browman and Goldstein's Articulatory Phonology (e.g., Browman & Goldstein, 1990) or "by hand". Thus, activation trajectories were modeled as switching discretely between values of zero (the gesture has no influence on tract shape) and one (the gesture has maximal influence on tract shape). However, it had been noted for some time that a simple step-function activation waveshape is an oversimplification (e.g., Bullock & Grossberg 1988; Coker 1976; Kröger, Schröder, & Opgen-Rhein 1995; Ostry, Gribble, & Gracco 1996), and we have since explored some of the consequences of non-step-function activation waveshapes on articulator kinematics. In particular, by explicitly specifying activation functions by hand to have half-cosine-shaped rises and falls of varying durations, we have been able to create articulatory trajectories whose kinematics capture individual differences in gestural velocity profiles found in experimental data (Byrd & Saltzman, 1998).

We have also explored two types of dynamical system for shaping activation trajectories in a relatively self-organized manner. Both types can be related to a class of rather generic recurrent connectionist network architectures (e.g., Jordan 1986, 1990, 1992; see also Bailly, Laboissière, & Schwartz, 1991; Lathroum, 1989). In such networks (see Figure 4), outputs correspond to gestural activations with one output node per gesture. The temporal patterning of gestural activation trajectories can, to a large extent, be viewed as the result of the network's state unit activity. This activity can be conceived as defining a dynamical flow with a time scale that is intrinsic to the intended speech sequence and that creates a temporal context within which gestural events can be located. The patterning of

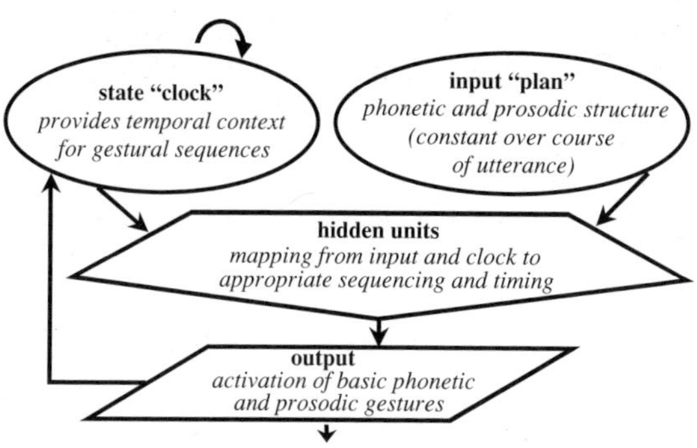

FIGURE 4. Trainable recurrent connectionist network that defines a dynamical system for patterning gestural activation trajectories.

activation trajectories is a consequence of the trained nonlinear mapping from the state unit flow to the output units' gestural activations. (For purposes of the present discussion we will ignore the plan units, which provide a unique identifying label for each sequence that the network is trained to perform, and that remain constant during the learning and performance of their associated sequences.)

There are two types of state unit structures that have been used in such networks (e.g., Jordan, 1986, 1992). In the first, each state unit is a self-recurrent, first-order filter that provides a decaying exponential representation of time, and that receives recurrent input from a given output unit. Thus, state units and output units are defined in a one-to-one manner, and the state units effectively provide a sequence-specific, exponentially weighted average of the activities of their associated output units. This is the type of state unit structure used in our previous work on the dynamics that give rise to a given gesture's anticipatory interval of coarticulation, defined operationally as the time from gestural motion onset to the time of required target attainment (e.g., Saltzman & Mitra, 1998; Saltzman, Löfqvist, & Mitra, 2000). Using such a model we were able to capture individual differences demonstrated experimentally among speakers in the temporal elasticity of anticipation intervals (e.g., Abry & Lallouache, 1995), according to which an interval lengthens (i.e., begins earlier), but only fractionally, with increasing numbers of preceding non-conflicting gestures.

The second type of state unit structure is a linear second-order filter composed of an antisymmetrically coupled pair of first-order units. Each such structure provides an oscillatory representation of time, generating a circular trajectory in the cartesian activity space of its two component units. The rate of angular rotation about this circle is a function of the weights of the units' self-recurrent and cross-coupling connections. In some cases, only one such oscillator is defined for a network with multiple output units (e.g., Bailly, Laboissière, & Schwartz, 1991; Laboissière, Schwartz, & Bailly, 1991; Jordan, 1990), and the state unit oscillator defines a relatively simple "clock", whereby the output units are activated whenever the clock passes through a corresponding set of time or phase values that are acquired during network training. For example, if two outputs correspond, respectively, to the activation of bilabial and laryngeal gestures for /p/, the relative phasing of these gestures would be determined by the relative values of their associated phase values in the state clock. Additionally, if the clock were gradually sped up in order to increase speaking rate, the gestural production rate would also change gradually along with the flow rate of the state clock, but intergestural relative phasing would not change.

One problem with this simple state clock model is that stability is lost in systematic ways when speaking rate is increased, such that abrupt shifts of intergestural relative phasing are observed at critical rates. Specifically, intergestural phase transitions are observed during rate-scaling experiments, showing discontinuous transitions of intergestural phasing with continuous increases in speaking rate (Kelso, Saltzman, & Tuller, 1986a, 1986b; Tuller & Kelso, 1991). In these studies, when subjects speak the syllable /pi/ repetitively at increasing rates, the relative phasing of the bilabial and laryngeal gestures associated with the /p/ does not change from the pattern observed at a self-selected, comfortable rate. However, when the repeated syllable /ip/ is similarly increased in rate, its relative phasing pattern switches relatively abruptly at a critical speed—from that observed for a self-selected, comfortable rate to the pattern observed for the /pi/ sequences. This phase transition in the intergestural timing of bilabial and laryngeal gestures implies that at least two separate state-unit oscillators exist, possibly one oscillator for each gestural unit, and that during the performance of a given sequence these state-oscillators behave as functionally coupled, nonlinear, limit-cycle oscillators. In unperturbed cases, the observed pattern of gestural activations would correspond to an associated pattern of synchronization (entrainment) and relative phasing among the state-oscillators that was acquired during training. Similarly, the intergestural phase transitions may be viewed as behaviors of a system of nonlinearly coupled, limit-cycle oscillators that bifurcate from a modal pattern that becomes unstable with increasing rate to another modal pattern that retains its stability (e.g., Haken, Kelso, & Bunz, 1985). The implications for models of activation dynamics are that the state unit clock should contain at least at least one oscillator per gesture, where the oscillators are governed by (nonlinear) limit cycle dynamics, and that these oscillators should be mutually coupled with one another.

The hypothesis of an ensemble of oscillators, one oscillator per gesture, that comprises a "clock" governing the timing of each gesture in a speech sequence echoes an earlier hypothesis by Browman and Goldstein (1990). According to their hypothesis, there is an abstract (linear) "timing oscillator" associated with each gestural unit, and that these oscillators are coupled in a manner that is responsible for the relative timing of gestural onsets and offsets in the sequence. Further, they hypothesized that different types of intergestural coupling structures existed within different parts of syllables and between syllables, and that timing patterns observed experimentally, both intra- and inter-syllabically, could be understood with reference to these coupling patterns. In the following section, we describe our recent work in which state-unit limit cycle oscillators ("planning oscillators") are defined in a 1:1 manner for each gestural activation node. In this work, a system graph is used to specify the coupling structure among the oscillators (i.e., the presence or absence of inter-oscillator linkages and, if present, the strength and target relative phase associated with each linkage) for a given gestural sequence, and the steady-state output of this oscillatory ensemble is used to specify the onsets and offsets of gestural activations for use in a gestural score for the sequence. Remarkably, the resultant relative timing patterns reflect both the mean values and variability observed experimentally for intra- and inter-syllabic intergestural timing patterns.

III. *Coupling Graphs and Intergestural Cohesion: Intra- and Inter-Syllabic Effects*

In our present work, we have extended Saltzman & Byrd's (2000) task dynamic model of intergestural phasing in a coupled pair of oscillators to the case in which multiple (more than two) oscillators are allowed to interact in shaping the steady-state pattern of intergestural phase differences (Nam & Saltzman, 2003). For a single pair of oscillators in the absence of added perturbations or noise, the system always settles or "relaxes" from its initial conditions (i.e., initial amplitude and phase for each oscillator) to a steady-state attractor characterized by a relative phasing pattern that is identical to the target relative phase value that is used to parameterize the bidirectional coupling function between the oscillators. When the system contains more than two oscillators, this is no longer necessarily the case, since the pairwise target relative phasings may be incompatible and in competition with one another. For example, for a system of three oscillators (A, B, C) with competing or incompatible target parameters for relative phase, e.g., 20° for the AB pair, 40° for the BC pair, and 30° for the AC pair, and with relatively equal strengths for each of the coupling functions (coupling strength is a second parameter of the coupling functions), none of the oscillator pairs will attain a steady-state relative phasing that matches the corresponding targets. For systems with unequal coupling strengths across the oscillator pairs, however, those pairs with relatively larger coupling strengths will attain steady-state relative phases closer to their targets than pairs with lesser coupling strengths. On the other hand, in a similar system with equal coupling strengths but with no such phasing target competition, e.g., 20° for the AB pair, 40° for the BC pair, and 60° for the AC pair, all oscillator pairs will achieve their targets in the steady-state.

Thus, the choice of which gestures to couple to one another (identified by nonzero-strength internode links in the oscillatory ensemble's system graph), as well as the relative strengths of the intergestural coupling functions, strongly influences the resultant steady-state patterns of intergestural timing. When we implemented the system graphs proposed by Browman & Goldstein (2000), we found that the model automatically displayed asymmetries of intrasyllabic gestural behavior that have been observed empirically, namely, that syllable-initial consonant sequences (onsets) behave differently from syllable-final sequences (codas) in two ways[1]. Onsets have been shown to display a characteristic pattern of mean intergestural relative phasing values, labeled the *C-center* effect by Browman and Goldstein (2000), that codas do not display. Additionally, onsets also exhibit less variability (i.e., greater stability) of intergestural relative phasing compared to the variability found in codas (Byrd, 1996).

The C-center effect describes the fact that, as consonants are added to onsets, the resultant timing of all consonant gestures changes with respect to the

[1] The *onset* of a syllable denotes the consonants in the syllable that precede the vowel; syllable *coda* denotes the consonants in a syllable following a vowel; syllable *rime (or rhyme)* denotes the vowel or *nucleus* of a syllable together with the following consonants in that syllable. So, for example, in the word *spritz*, *spr* is the onset; *i* is the vocalic nucleus, *tz* is the coda; and *itz* is the rime.

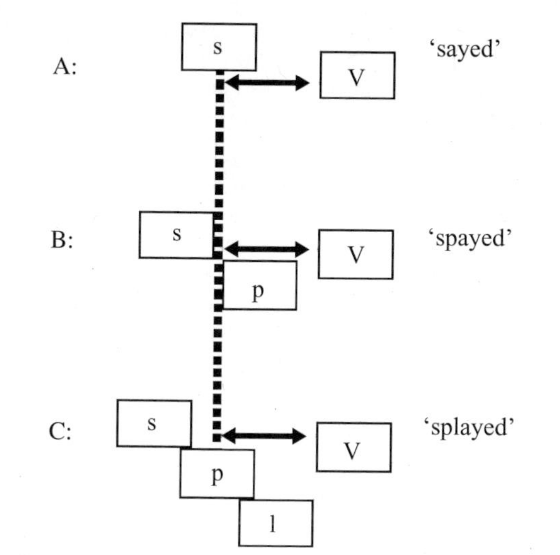

FIGURE 5. Articulatory gestural schematics derived from X-ray micro-beam data in Browman and Goldstein (2000) for consonant-vowel sequences in 'sayed', 'spayed' and 'splayed'. Vertical dotted line denotes the temporal "centers of gravity" (C-centers) for the onset consonant gestures. Horizontal arrows show invariant time from onset centers to the vowels (C-center effect).

following vowel in a way that preserves the overall timing of the *center* of the consonant sequence with respect to the vowel (see Figure 5). In contrast, however, as consonants are added to coda sequences, the temporal distance of the center of the cluster from the preceding vowel simply increases with the number of added consonants. Browman & Goldstein (2000) hypothesized that these different behaviors originated in different underlying coupling structures for the component gestures in onsets and codas. As shown in Figure 6, these different structures can be represented as correspondingly different system graphs. The graph

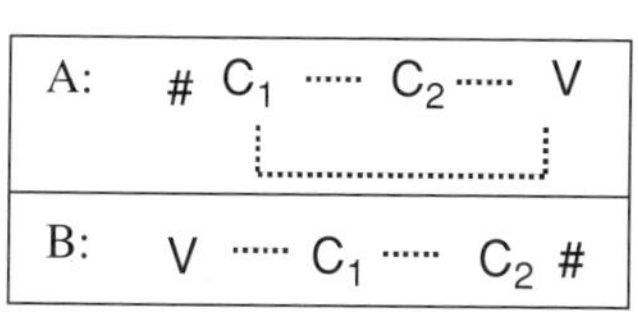

FIGURE 6. Coupling graphs proposed by Browman & Goldstein (2000) for syllable onsets (6A) and codas (6B). #s denote syllable boundaries.

for onsets (Figure 6A) defines the C-C coupling as well as (identical) C-V couplings for each consonant to the vowel, and there is competitive interaction between the C-C and C-V couplings; for codas (Figure 6B), however, the graph defines a similar C-C coupling, but only the first consonant is coupled to the preceding vowel (V-C coupling) and there is no comparable competition.

We used these onset and coda coupling graphs to parameterize simulations based on our extended coupled oscillator model consisting of three pairwise-coupled oscillators (Nam & Saltzman, 2003). Implementing the graph in Figure 6A for onset cluster simulations, we set the target relative phase parameters of the coupling functions to 50°, 50° and 30°, respectively, for the C_1-V, C_2-V couplings (Figure 7, top row, left) and the C_1-C_2 couplings (Figure 7, top row, right). All coupling strength parameters were set to equal 1. When the system settled into its entrained steady-state, the resultant intergestural relative phases were 59.94° for C_1-V, 39.96° for C_2-V, and 19.98° for C_1-C_2 (Figure 7, bottom). Thus, implementing the graph in Figure 6A resulted in none of the intergestural relative phases achieving their target values due to the competitive interactions between the C-V and C-C couplings. Importantly, however, the

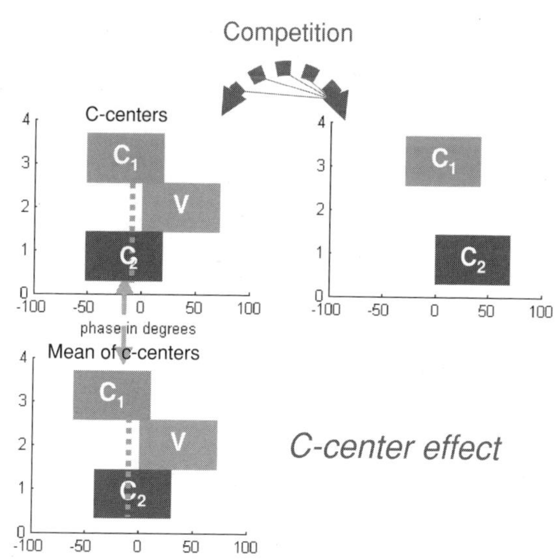

FIGURE 7. C-center effect in CCV. Top row, left: Target relative phases for C1-V and C2-V = 50°; Top row, right: C1-C2 target relative phase = 30°; Bottom row: CV and CC phasings are in competition and result in a C-center effect.

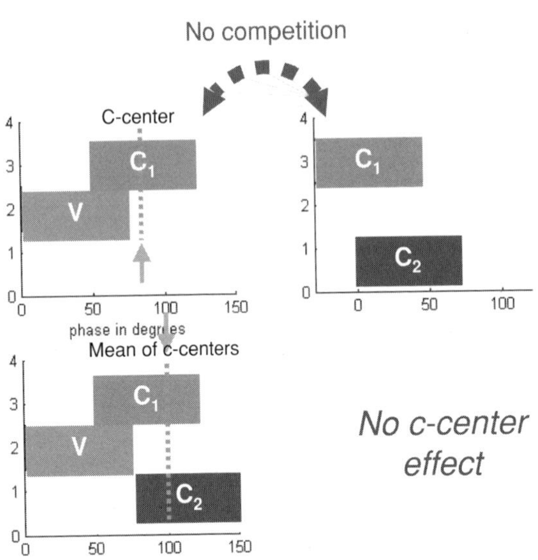

FIGURE 8. No C-center effect in VCC. Top row, left: Target relative phase for V-C_1 = 50° but is unspecified for V-C_2; Top row, right: C1-C2 target relative phase = 30°; Lack of competition between CV and CC phasings result in absence of C-center effect (Bottom row).

mean phase of the C_1C_2 onset consonant cluster displayed the C-center effect (see the vertical dotted line in the left column of Figure 7), and the mean phase of C_1 and C_2 relative to the vowel was equal to the target CV phasing of 50° (i.e., [60° + 40°] / 2) despite the "failure" of each consonant to sustain the target relative phase (50°) with respect to the vowel due to the competition.

For simulating coda clusters we implemented the graphs shown in Figure 6B, setting the target relative phases of the V-C_1 and C_1-C_2 pairs to 50° and 30°, respectively (Figure 8, top row). No coupling was specified for the V-C_2 pair and, thus, there was no competition between coupling terms in this case. Again, all coupling strengths were set to equal 1. When these coda simulations settled into steady-state, the final relative phasings between the gestural oscillators were 49.96°, 29.94°, and 79.90° for the V-C_1, C_1-C_2, and V-C_2 pairs, respectively (Figure 8, bottom row). Note that each of two targeted phase relations is achieved and the resultant relative phase of V-C_2 is the simple sum of the relative phases of V-C_1 and C_1-C_2. Thus, the resultant relative phase between the vowel and the mean phase of C_1 and C_2 is 65°, which is different from the target VC relative phase (50°), and there is no C-center effect.

The above simulations focused on the effects of system graph structure on patterns of intergestural relative phasing displayed within syllables. These intrasyllabic simulations were purely deterministic and contained no contributions of stochasticity or noise. Consequently, the steady-state patterns observed for a given parameterization and set of initial conditions behaved identically from one "trial" to the next, and may best be considered to reflect the mean values of intergestural phasing observed experimentally. There is also, however, a behavioral asymmetry in the relative stability or variability found in onsets and codas—intergestural phasing is less variable (more stable) in onsets than in codas (Byrd, 1996). Browman & Goldstein (2000) hypothesized that, similar to the asymmetries found between mean intergestural phasing patterns, the asymmetries in stability could also be accounted for by the differences between coupling graphs for onsets (Figure 6A) and codas (Figure 6B). To test this hypothesis we conducted a set of simulations in which stochasticity was incorporated and the resultant variability of relative phasing was measured.

In these simulations, we used the same onset and coda graphs as above (Figure 6A and 6B), and incorporated variability by introducing trial-to-trial random variation in the *detuning* (i.e., the difference between the natural frequencies) of the component oscillator pairs. We ran groups of simulated trials for each utterance type, adding a random amount of detuning in each trial to each oscillator pair via the associated interoscillator coupling function. The detuning parameter, b, in each coupling function was defined as a random variable with a mean and standard deviation equal to zero and σ, respectively. The value of σ (the amount of detuning noise) was manipulated in a series of five noise conditions, increasing from .05 to .85 in .20 increments. 200 simulation trials were run for each utterance (onset, coda) x noise-level (5 levels) condition, and we measured the standard deviation of the final steady-state relative phase between C_1 and C_2 for each condition. Figure 9 displays the amount of variability shown in each condition. Not surprisingly, for both onset and consonant clusters there is greater resultant steady-state variability as the added noise level increases. More importantly, however, is the fact that at each noise level the variablility in relative phasing is smaller for onset clusters than for coda clusters. This reflects the experimental finding that onset clusters are more stable than coda clusters in their relative timing (Byrd, 1996), and supports the hypothesis of Browman and Goldstein (2000) that this asymmetry in variability/stability is the emergent consequence of differences between onsets and codas in their underlying intergestural coupling graphs.

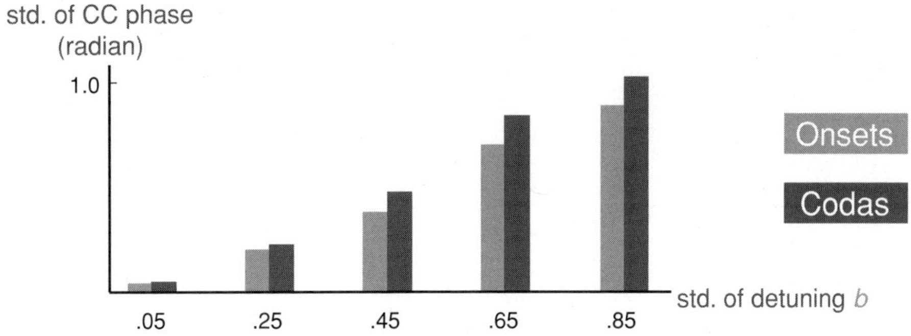

FIGURE 9. Standard deviation of steady-state relative phasing for onset and coda consonant clusters across five levels of input noise.

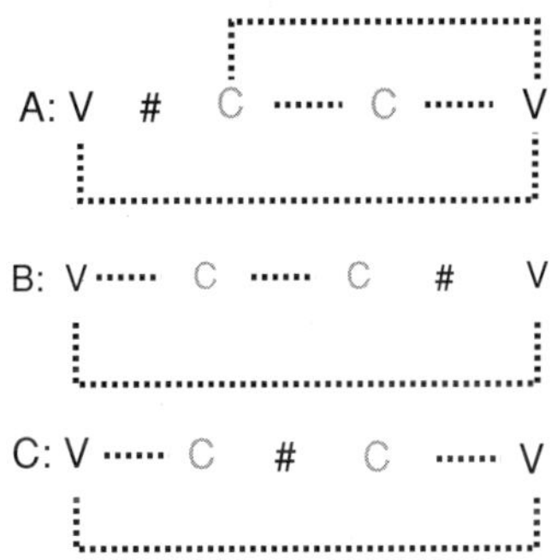

FIGURE 10. Coupling graphs for syllable sequences. Consonant clusters are in onset position (A), coda position (B), or span the syllable boundary (C). #s denote syllable boundaries.

Encouraged by these results, we extended our model to four-gesture sequences that were defined across syllable (or word) boundaries. In terms of the underlying coupling graphs, we made the rather minimal assumption that only vowels (i.e., syllable nuclei) are coupled across syllable boundaries, and compared C-C variability in relative phasing for three types of simulated sequences: onset consonant clusters, coda clusters, and clusters spanning the syllable boundary. Figure 10 illustrates the graphs used in these simulations. As with our intrasyllabic modeling, we included the same five levels of detuning noise; we then measured the standard deviation of the final steady-state C-C relative phase for 200 trials of each utterance (onset, coda, cross-syllable), x noise-level condition. The results are shown in Figure 11. Again, not surprisingly, the level of resultant steady-state variability of intergestural phasing reflects the level of input noise for all

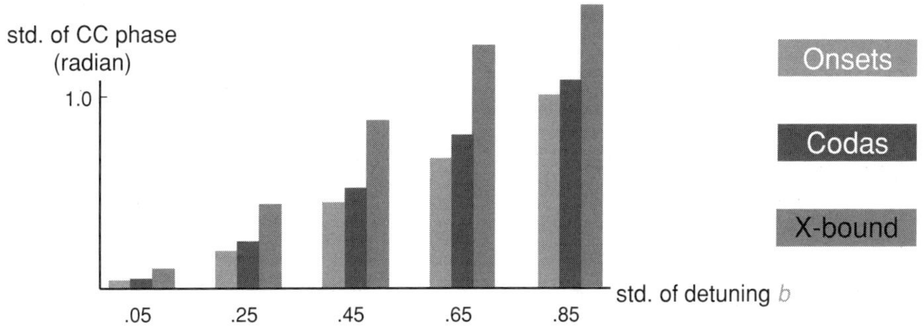

FIGURE 11. Standard deviation of steady-state relative phasing for consonant clusters in onsets, codas, and spanning syllable boundaries across five levels of input noise.

utterance conditions. Additionally, the variability in onset clusters in smaller than in coda clusters, replicating our earlier results but now using larger gestural ensembles that include intersyllabic coupling. Finally, and crucially, the variability across syllabic boundaries was largest of all, in agreement with empirical data reported by Byrd (1996).

IV. *Concluding Remarks*

We have reviewed the manner in which dynamical systems for coordination and control can be analyzed in terms of their state variables, parameters, and graphs. For the most part, system graphs have been ignored in studies focusing on the spatiotemporal properties of skilled actions. However, at least in the case of intergestural timing patterns in speech production, it appears that system graphs can be invoked not only as the source of the mean timing properties of an utterance, but of the particular structure of its variability as well. We are encouraged by the power of this approach and are both curious and eager to see how this focus can be brought to bear as well on the study of nonspeech skilled activities.

Acknowledgements

This work was supported by NIH grants DC-03663 and DC-03172.

References

Abry, C. & Lallouache, T. (1995). Modeling lip constriction anticipatory behavior for rounding in French with MEM (Movement Expansion Model). In K. Elenius & P. Branderud, (Eds.). *Proceedings of the XIIth International Congress of Phonetic Sciences.* Stockholm: KTH and Stockholm University, pp. 152–155.

Bailly, G., Laboissière, R., & Schwartz, J. L. (1991). Formant trajectories as audible gestures: An alternative for speech synthesis. *Journal of Phonetics, 19,* 9–23.

Browman, C. P., & Goldstein, L. (1990). Tiers in articulatory phonology, with some implications for casual speech. In J. Kingston & M. E. Beckman (Eds.), *Papers in laboratory phonology: I. Between the grammar and the physics of speech.* Cambridge, England: Cambridge University Press. Pp. 341–338.

Browman, C. P., & Goldstein, L. (1995). Gestural syllable position effects in American English. In F. Bell-Berti & L. Raphael, (Eds.). *Producing speech: Contemporary issues.* Woodbury, New York: American Institute of Physics. Pp. 19–33.

Browman, C. P. & Goldstein, L. Competing constraints on intergestural coordination and self-organization of phonological structures. *Bulletin de la Communication Parlée, 5*: 25–34, 2000.

Bullock, D., & Grossberg, S. (1988). Neural dynamics of planned arm movements: Emergent invariants and speed-accuracy properties during trajectory formation. *Psychological Review,* 95, 49–90.

Byrd, D. (1996) Influences on articulatory timing in consonant sequences. *Journal of Phonetics, 24*(2), 209–244, 1996.

Byrd D. & Saltzman, E. (1998) Intragestural dynamics of multiple prosodic boundaries. *Journal of Phonetics, 26,* 173–199.

Coker, C. H. (1976). A model of articulatory dynamics and control. *Proceedings of the IEEE, 64,* 452–460.

Farmer, J. D. (1990). A Rosetta Stone for connectionism. *Physica D, 42,* 153–187.

Huang, G.-B., Saratchandran, P., & Sundararajan, N. (2005). A generalized growing and pruning RBF (GGAP-RBF) neural network for function approximation. *IEEE Transactions on Neural Networks, 16* (1), 57–67.

Jordan M.I. (1990) Motor learning and the degrees of freedom problem. In Jeannerod M, (ed) *Attention and Performance XIII.* Hillsdale, NJ: Erlbaum.

Jordan M.I. (1992) Constrained supervised learning. *Journal of Mathematical Psychology,* 36, 396–425.

Jordan, M. I., & Rumelhart, D. E. (1992). Forward models: Supervised learning with a distal teacher. *Cognitive Science, 16,* 307–354.

Kelso, J. A. S., Saltzman, E. L., & Tuller, B. (1986a). The dynamical perspective on speech production: Data and theory. *Journal of Phonetics, 14,* 29–60.

Kelso, J. A. S., Saltzman, E. L., & Tuller, B. (1986b). Intentional contents, communicative context, and task dynamics: A reply to the commentators. *Journal of Phonetics, 14,* 171–196.

Kröger, B., Schröder, G. and Opgen-Rhein, C. (1995) A gesture-based dynamic model describing articulatory movement data. *Journal of the Acoustical Society of America 98.4* 1878–1889.

Laboissière, R., Schwartz, J.-L. & Bailly, G. Motor control for speech skills: A connectionist approach. *Connectionist models. Proceedings of the 1990 Summer School.* (D. S. Touretzky, J. L. Elman, T. J. Sejnowski & G. E. Hinton, editors), pp. 319–327. San Mateo, CA: Morgan Kaufmann, 1991.

Lathroum, A. (1989). Feature encoding by neural nets. *Phonology, 6,* 305–316.

Nam, H., & Saltzman, E. (2003). A competitive, coupled oscillator model of syllable structure. In *Proceedings of the 15th International Congress of Phonetic Sciences (ICPhS-15),* Barcelona, Spain, 2003.

Ostry, D. J., Gribble, P. and Gracco, V. L. (1996) Coarticulation of jaw movements in speech production: Is context sensitivity in speech kinematics centrally planned? *The Journal of Neuroscience, 16*(4), 1570–1579.

Quartz, S. R., & Sejnowski, T. J. (1997). The neural base of cognitive development: A constructivist manifesto. *Behavioral and Brain Sciences, 20,* 537–596.

Saltzman, E. (1986). Task dynamic coordination of the speech articulators: A preliminary model. *Experimental Brain Research, Series 15,* 129–144.

Saltzman, E., & Byrd, D. (2000). Task dynamics of gestural timing: Phase windows and multifrequency rhythms. *Human Movement Science, 19,* 499–526.

Saltzman, E., Löfqvist, A., & Mitra, S. (2000). "Glue" and "clocks": Intergestural cohesion and global timing. In *Papers in Laboratory Phonology V*. (M. B. Broe & J. B. Pierrehumbert, editors), pp. 88–101. Cambridge: Cambridge University Press.

Saltzman, E., & Mitra, S. (1998). A task-dynamic approach to gestural patterning in speech: A hybrid recurrent network. *Journal of the Acoustical Society of America, 103,* (5; Pt.2), 2893 (Abstract).

Saltzman, E. L., & Munhall, K. G. (1989). A dynamical approach to gestural patterning in speech production. *Ecological Psychology, 1,* 333–382.

Saltzman, E., & Munhall, K. G. (1992). Skill acquisition and development: The roles of state-, parameter-, and graph-dynamics. *Journal of Motor Behavior, 24(1),* 49–57.

Tuller, B. & Kelso, J. A. S. (1991). The production and perception of syllable structure. *Journal of Speech and Hearing Research, 34,* 501–508.

III. MOTOR LEARNING AND NEURAL PLASTICITY

7. STABILIZATION OF OLD AND NEW POSTURAL PATTERNS IN STANDING HUMANS

Benoît G. Bardy

University of Montpellier-1, France; Institut Universitaire de France

Elise Faugloire

School of Kinesiology, University of Minnesota, MN, USA

Paul Fourcade

University of Paris 11, Orsay, France

Thomas A. Stoffregen

School of Kinesiology, University of Minnesota, MN, USA

Abstract

In human stance, rotations around the hips and ankles typically exhibit a relative phase close to 20°, or close to 180°. In this article, we propose a model of stance that captures these postural states and the changes between them. We also describe the results of a recent study in which participants learned a novel pattern of hip and ankle coordination (a relative phase of 135°). Participants learned this novel pattern rapidly. At the same time, learning led to a robust destabilization of pre-existing patterns of hip-ankle coordination. The rate and type of destabilization depended upon the initial stability of the pre-existing patterns. We discuss similarities and differences between the learning of postural and bimanual coordination modes.

Stabilization of Old and New Postural Patterns in Standing Humans

The maintenance of stable upright stance is required in many daily activities in humans and other bipeds. Stable stance requires the body's center of mass to be kept above the feet. Postural control actions consist mainly of coordinated rotations around the hips and ankles. Many patterns of ankle-hip coordination will maintain the center of mass above the feet, but only a few of these are effective across a broad range of situations. Functionality depends in part on the task in which the person is engaged. For example, some coordination patterns that prevent falling may be avoided because they hamper the realization of other, simultaneous goals, such as maintaining gaze, or manual contact with an object. Other coordination patterns may both prevent falling and facilitate performance on these *supra-postural* tasks, and so may be preferred (e.g., Bardy, Marin, Stoffregen, & Bootsma, 1999; Stoffregen, Smart, Bardy, & Pagulayan, 1999). This fact has implications for pre-existing patterns of postural coordination, but also for the acquisition of new patterns. In the present contribution, we use existing data on postural transitions to develop a simple model of postural states and postural changes that allow the realization of supra-postural activities. We then examine the problem of learning new postural coordination patterns within the framework of non-linear dynamics of perception and action.

POSTURAL PERSISTENCE AND POSTURAL CHANGE

In our research on postural dynamics (e.g., Bardy et al., 1999; Bardy et al., 2002; Marin et al., 1999; Oullier et al., 2002, 2004), standing participants have been instructed to maintain a constant distance between their head and a visual target that oscillates along the line of sight. They have not been given any instructions about how standing posture was to be controlled during the tracking task. We measured rotations at the ankles and hips, and analyzed the relative phase, ϕ_{rel}, of rotations at these joints. Two coordination modes between ankles and hips have been consistently observed. An *in-phase* mode, with ϕ_{rel} of about 20–25°, emerged when the visual tracking target moved at small amplitude (e.g., Bardy et al., 1999) or low frequency (e.g., Bardy et al., 2002). An *anti-phase* mode, with ϕ_{rel} close to 180°, has emerged when the visual target moved with large amplitude or high frequency. The departure from pure in-phase motion ($\phi_{rel} = 0°$) found for low amplitude and frequency contrasts with studies of bimanual coordination (e.g., Haken, Kelso, & Bunz, 1985; Kelso, 1984) and may be a consequence of the frequency competition, $\Delta\omega$, between the upper and lower parts of the body (e.g., Sternad, Amazeen, & Turvey, 1996). It might also result from the mechanical constraint of maintaining the center of mass above the base of support. The differential emergence of these modes was influenced by intentional constraints (i.e., the instruction to track target motion), by behavioral constraints (i.e., height of the center of mass, length of the feet, body stiffness, expertise in sport), and by environmental constraints (i.e., surface properties, target amplitude or frequency); (see Bardy, 2004 for a review). It was the simultaneous, interacting pressures—cooperative or competitive—imposed by the task, the body, and the environment that determined the selective emergence of the in-phase and anti-phase modes (cf. Newell, 1986).

We also observed that transitions between in-phase and anti-phase ankle-hip modes revealed characteristics of *non-equilibrium phase transitions* (Bardy et al., 2002). As we increased or decreased the frequency at which the visual target moved, a frequency-induced loss of stability occurred, yielding *critical fluctuations* in the vicinity of the region of the frequency range in which there was a transition between coordination patterns (see Figure 1). Transitions between in-phase and anti-phase modes were abrupt, and exhibited *hysteresis*: Transitions from in-phase to anti-phase occurred at a higher frequency of target motion than transitions from anti-phase to in-phase. Finally, we applied an external perturbation (a sudden shift in the direction in which the target was moving). The perturbation was applied either near to or far from the region (frequencies) in which transitions between modes were known to occur. Each mode was found to be less stable when the perturbation was applied close to the transition region, and more stable when it was applied far from it, as evidenced by a larger relaxation time in the latter situation (*critical slowing down*). In summary, our research has shown that postural modes (i) emerge out of the coalescence of multiple constraints, (ii) exhibit persistence and change that are characteristic of self-organized systems, and (iii) are modulated by the actor's intentions. In the next two sections, we review evidence that (iv) the postural behavior can be modeled using simple tools from biomechanics, and that (v) the learning

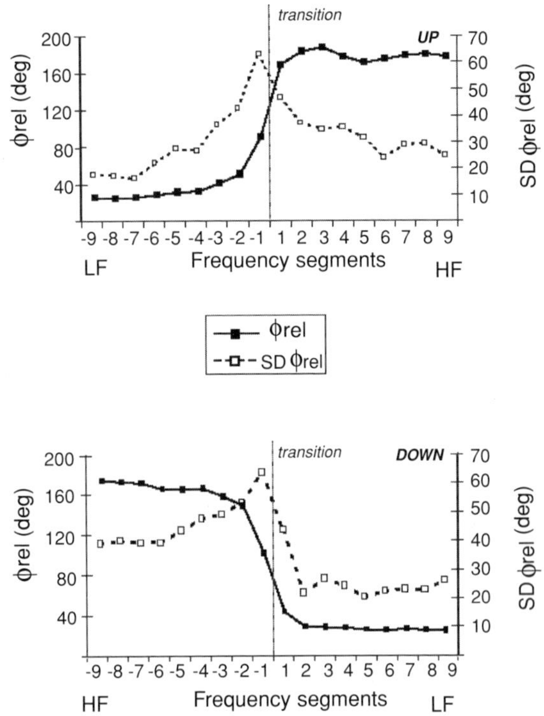

FIGURE 1. Postural transitions in a visual tracking task: Mean point estimate ankle-hip relative phase ϕ_{rel} (and *SD*) in *Up* and *Down* conditions (10 participants). Each segment includes a temporal average of ϕ_{rel} over 4 cycles of oscillation, with an overlap of two cycles. *LF* and *HF* refer to low frequency and high frequency segments respectively. Adapted from Bardy et al. (2002). The dynamics of human postural transitions. *Journal of Experimental Psychology: Human Perception and Performance*, 28, 499–514.

of new postural modes is accompanied by destabilization of pre-existing dynamics of the postural system.

A SIMPLE MECHANICAL MODEL OF POSTURAL PERSISTENCE AND POSTURAL CHANGE

Research on posture has been a fertile ground for the development of structural or phenomenological models. In pioneering work, Nashner (1976; Nashner & McCollum, 1985), proposed that postural coordination in the maintenance of stance is rooted in the organization of the neuromuscular system. Since then, a variety of mechanical or neurophysiological, structure-related models of the postural system have been proposed (e.g., Barin, 1989; Kuo, 1995; Stockwell, Koozekanani & Barin 1981; Yang, Winter & Wells, 1990). In a different meta-theoretical context, recent developments in the non-linear dynamics of perception and action have inspired the emergence of phenomenological, structure-free models accounting for the self-organization of posture and gait. These latter models use mathematical tools borrowed from non-equilibrium statistical mechanics and stochastic dynamics (e.g., Balasubramaniam, Riley & Turvey, 2000; Dikjstra, Schöner & Gielen, 1994; Kay & Warren, 2001; Schöner, 1991). Attempts to mix the structure-related and structure free models are rare, but do exist. For instance, Taga (1994, 1995) proposed a multi-level model of human locomotion in which the coordination dynamics observable at the behavioral level (i.e., the gait) are consequences of interactions between the neural system and the musculo-skeletal system. Here we propose a simple, mixed model of human posture that captures the behavior of the postural system that has been observed in the tracking task described earlier. Our model is biologically plausible, and is composite in the sense that it is a mechanical model that links joints and segments, with units of mass and length.

The Model. Our preliminary model is a simple, two-segment inverted pendulum system representing the human body. The upper segment represents the head-arms-trunk system and the lower segment represents the legs (*cf.* Figure 2). The segment masses are concentrated and localized at their centers of gravity, and are noted by m_1, m_2 for the trunk and legs, respectively. The segment lengths are noted by l_1, l_2 for trunk and legs, respectively. Muscular and articular damping and stiffness terms are present only at the level of the joints (ankle, hip). The motor command is modeled by the application of a constant torque at the ankle joint. This choice is appropriate when the amplitude of the tracked target oscillations is small (5 cm in the simulation). A second active torque operating at the hip joint, but with an opposite sign, compensates for the growing inertial force that accompanies an increase in movement frequency, thus maintaining the system in balance. The amplitude of the second torque is not constant but increases exponentially with frequency until an asymptotic value is reached in the vicinity of the transition zone.

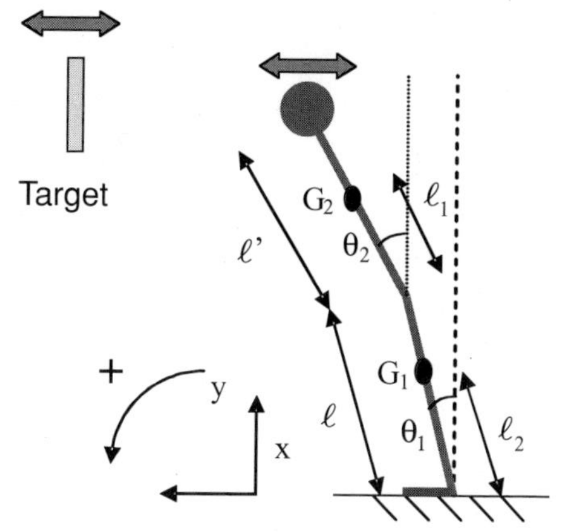

FIGURE 2. Double-inverted pendulum model of the human body during the simulated tracking task. Rotations of the two segments with respect to gravity are given by θ_1 (legs) and θ_2 (trunk). G_1 and G_2 refer to the position of the centers of mass of the two segments and are located at distances l_1 and l_2 from their axis of rotation. l and l' refer to the length of the two segments.

On the basis on these considerations, the differential equations for the motion of the system were computed, derived from Lagrange's equations:

$$\ddot{\theta}_1 + f_1(\dot{\theta}_1) + g_1(\theta_1) = I_1(\dot{\theta}_1, \theta_1, \dot{\theta}_2, \theta_2)$$
$$\ddot{\theta}_2 + f_2(\dot{\theta}_2) + g_2(\theta_2) = I_2(\dot{\theta}_1, \theta_1, \dot{\theta}_2, \theta_2)$$

where f_i (i = 1,2) is the damping function, g_i (i = 1,2) is the stiffness function and I_i (i = 1,2) is the coupling function between the two joints.

Simulations. In order to match the target oscillation amplitude in the experimental studies summarized above, torque amplitude of 15 N.m was chosen for the ankles. As a result of this choice, the torque amplitude at the hip joint needed to be greater than 5 N.m but smaller than 20 N.m in order to counterbalance the inertial forces while maintaining the amplitude of the head. Stiffness coefficients acting at the joints were estimated at 1100 N.m.rad^{-1} for the ankles and

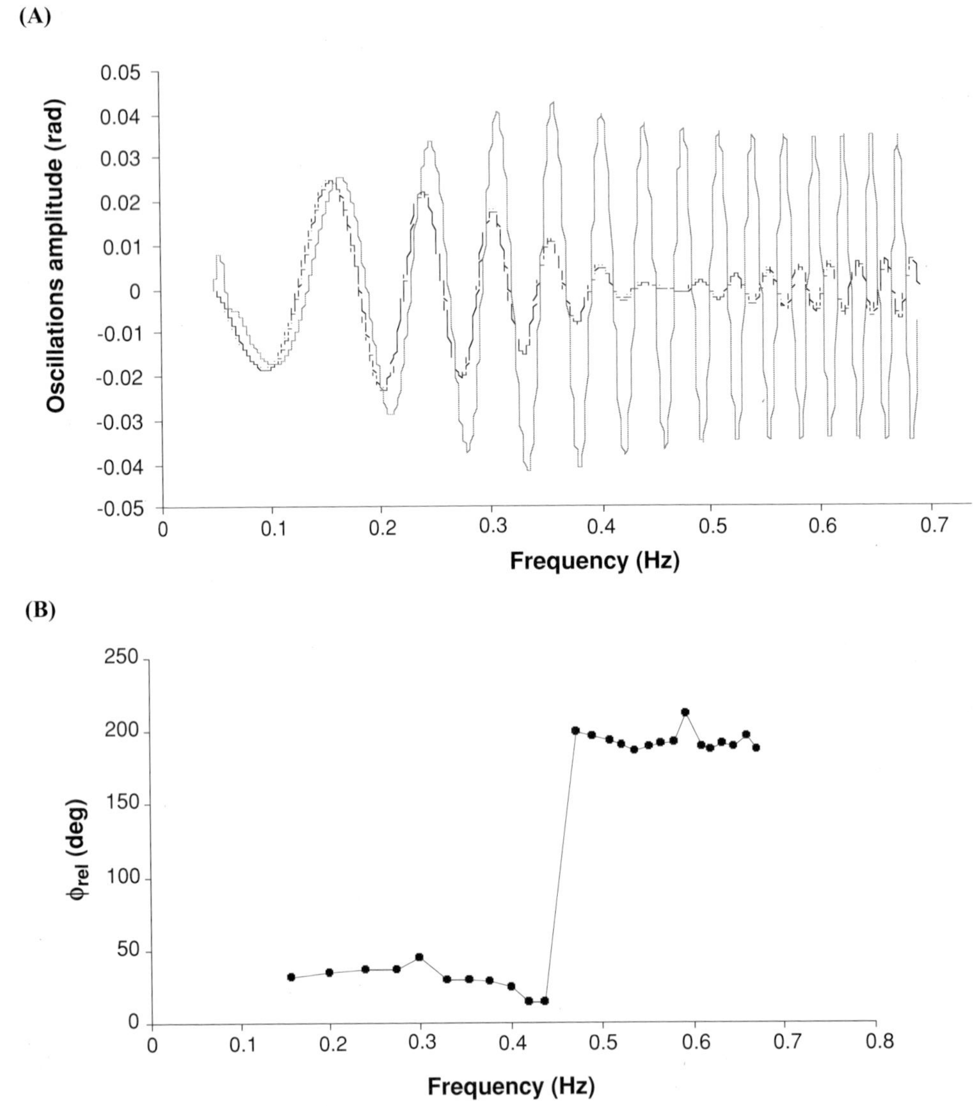

FIGURE 3. Transition region for one numerical simulation (Up condition), showing angular ankle and hip amplitudes (in radians, Figure 3a) as well as the (point-estimate) relative phase ϕ_{rel} (Figure 3b) as a function of target frequency. Sustained in phase motion at low frequency, anti-phase motion at high frequency, and a transition from an in-phase to anti-phase at $f = 0.45$ Hz can be observed. Parameters for this simulation were $\Gamma_1 = 15$ N.m, $\Gamma_2 = 10$ N.m, weight = 70 kg, size = 1.75 m.

300 N.m.rad^{-1} for the hips according to the literature (Farley & Morgenroth, 1999, Stefanyshyn & Nigg, 1998, Weiss, Hunter & Kearney, 1988). The simulations described below were obtained for a typical subject of intermediate height (175 cm) and weight (70 kg). Local masses and positions of local centers of were respectively provided by Winter (1990) and Le Veau (1977).

The main behavior resulting from the numerical simulations (performed with Matlab©) is shown in Figure 3, representing ankle and hip oscillations as a function of frequency. For low frequencies, the model exhibited an ankle-hip relative phase close to 0°, and an ankle-hip relative phase close to 180° for large frequency values. The system abruptly switched from an in-phase mode to an anti-phase mode at a specific frequency value (0.45 Hz in Figure 3). Another interesting outcome of the model is the decrease in amplitude that accompanied the increase in frequency, toward zero at the transition point, with a reverse increase in amplitude after the transition point (i.e., after emergence of the anti-phase mode). This typical behavior was induced by the segmental inertia of the coupled components.

This simple mechanical model of the human body captures two of the essential properties that have been observed in standing humans involved in suprapostural activities (e.g., Bardy et al., 2002): the presence of two attractors for the relative phase of ankle-hip coordination, and a frequency-induced transition between attractors as target frequency increases. The model has been developed to reproduce *critical fluctuations*, that is, the increase in the standard deviation of ϕ_{rel} in the vicinity of the transition, *hysteresis*, that is, the tendency of a system to remain in its current basin of attraction as a control parameter moves through the transition region, and *differential critical slowing down* far and close to the transition (Fourcade, Bardy & Roudeix, 2005). The success of this model suggests that it is possible, and we would say necessary, to root the general organizational principles accompanying movement control into the biomechanical (or neuro-physiological) substrates of specific biological systems, such as the postural system. It also suggests that an intermediate position on the structural-phenomenological line (c.f., Beek et al., 1998) can be a useful route to follow for modeling human movement.

THE LEARNING OF NEW POSTURAL PATTERNS

The experimental and modeling efforts reported above offer evidence for the existence of self-organization in whole-body coordination. They encourage further examination of the possibility that the interactions between the components of the postural system may be understood through the physics of non-equilibrium processes. In our initial study of transitions between postural coordination modes (Bardy et al., 2002) the two modes that we observed emerged spontaneously (i.e., without instruction). In this section, we examine a complementary question of our research agenda on postural dynamics, related to how new postural modes are learned (see Faugloire, 2005; for a detailed treatment). We suggest that learning a new mode of postural coordination depends heavily on the competition between the dynamics of the new, to-be-learned pattern and the dynamics of pre-existing, stable postural patterns. We briefly describe the results of a study in which participants learned a new multi-joint postural coordination (Bardy, Faugloire, & Stoffregen, 2005).

In bimanual coordination, Zanone and Kelso (e.g., 1992; 1997) have elaborated a dynamical account of motor learning, based on the fact that the process of learning a new coordination pattern interacts with pre-existing states of the motor system. This interaction consists mainly of two phenomena. First, preferred and stable coordination tendencies systematically affect the ability to learn a new pattern: The more stable the initial states, the more difficult the learning of the new pattern. Second, learning a new mode changes the entire dynamics of the motor system, and can destabilize pre-existing (and previously stable) modes. In bimanual coordination, the two facets of this interaction have been explored in several studies (e.g., Fontaine, Lee & Swinnen, 1997; Lee, Swinnen & Verschueren, 1995; Smethurst & Carson, 2001; Wenderoth & Bock, 2001; Zanone & Kelso, 1992; 1997; Kelso & Zanone, 2002). Both convergences and divergences between theoretical predictions and experimental results have been found. As far as we know, only bimanual systems (fingers or arms) have been used to address these questions, and in terms of theory-testing the literature is cruelly lacking data from other motor systems. Formally, the dynamics of learning a new coordination should follow the same principles, irrespective of the effector system involved. However, the postural system (PS) is very different from the bimanual system (BS), in terms of the number of degrees of freedom involved (few for BS, many for PS), the eigenfrequency of the components involved (identical for BS, different for PS), and the type of coupling (perceptual for BS, perceptual and inertial for PS). Thus, the postural system may be a good candidate to test the generality of a dynamic theory of learning.

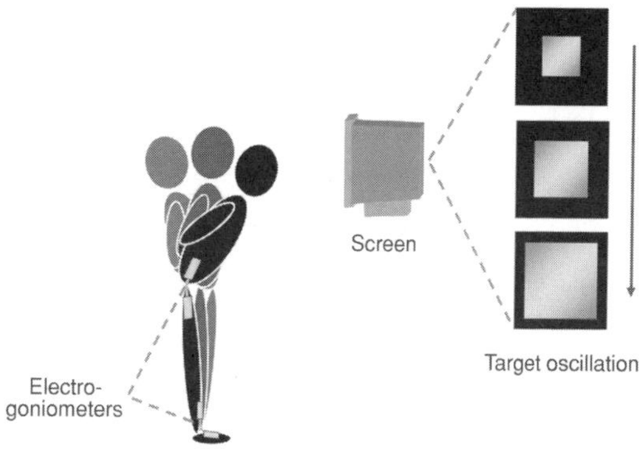

FIGURE 4. The visual tracking task used for the pre-test and post-test. Participants were instructed to follow with the head the antero-posterior oscillations of a moving target.

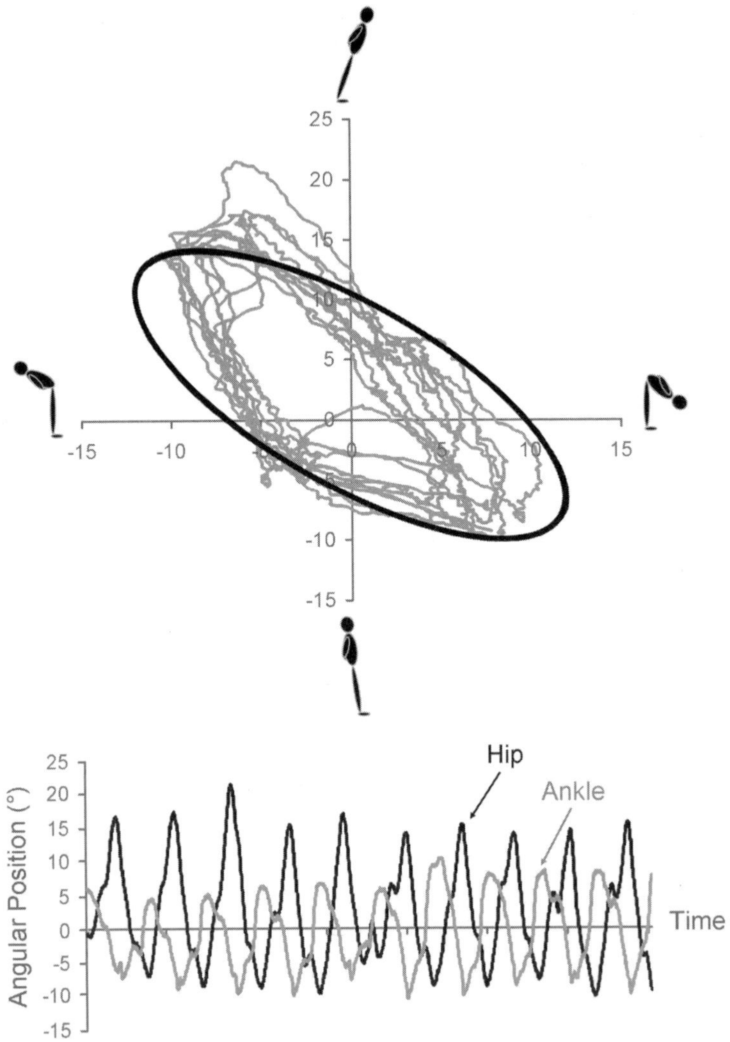

FIGURE 5. Top: Feed-back given to participants after every third trial showing the discrepancy in the ankle-hip plane (state space, or Lissajous plot) between the current pattern and the pattern to be learned (135°); Bottom: Hip and ankle movements (in degrees) over time during the production of a 135° relative phase pattern.

We recently carried out experiments in order to examine the learning of new postural patterns and its consequences on the stability of spontaneous, initial patterns (Bardy et al., 2005). We chose a relative phase of 135° between ankle and hip movements. A relative phase of 135° does not occur spontaneously in stance (so far as we know), has not been observed previously in our studies, and seems learnable. In one experiment, we investigated interactions between the process of learning the 135° pattern and the pre-existing anti-phase pattern. In our second experiment, we investigated how the process of learning the 135° pattern affected the pre-existing in-phase pattern. We also tested the durability of the pattern modifications, using a retention test conducted one week after the learning session.

Experiment 1 consisted of three sessions, using a pre-test/post-test design realized over two consecutive days. In the first session, or pre-test, participants ($N = 11$) performed the supra-postural task illustrated in Figure 4, with the instruction to maintain a constant distance between the head and a visible target that oscillated in the anterior-posterior axis (see Bardy et al., 1999 for details). They performed four trials of 10 oscillations each. We measured angular displacements of ankle and hip joints with two electrogoniometers connected to a DATALINK interface (Biometrics, Inc.). The emerging coordination in performing the supra-postural task was called the *initial spontaneous coordination*, and was characterized by the (discrete) ankle-hip relative phase, ϕ_{rel}. Its standard deviation, $SD\phi_{rel}$, indicated the stability of the coordination. The frequency and the amplitude of target motion were chosen to produce an anti-phase pattern (the in-phase pattern was investigated in Experiment 2). The second part of the experiment was the learning session. Participants attempted to learn the 135° relative phase based on explanations and demonstrations, but with no target to track. The learning phase consisted of 30 trials of 10 oscillations each, 15 on the first day, and 15 on the second day. After every third trials, participants were given feedback indicating the discrepancy between the performed coordination and the to-be-learned pattern (Figure 5). Finally, in a post-test after the learning phase, we repeated the pre-test tracking task to assess the effects of learning on the stability of initial spontaneous patterns. Therefore, participants performed the supra-postural task (tracking the target) during the pre-test and post-test sessions, in between which they attempted to learn the new 135° pattern.

Experiment 2 repeated the same design, with the following changes. First, to examine the effect of learning on both the in-phase and anti-phase patterns, different groups of participants were tested with low frequency ($N = 5$) and high frequency ($N = 6$) target motion. Based on previous studies using the supra-postural task (Bardy et al., 1999; 2002; Marin et al., 1999), we expected that the low target frequency (0.25 Hz) would induce an in-phase ankle-hip coordination while the high target frequency (0.65 Hz) would produce an anti-phase pattern. The second change compared to Experiment 1 was that the learning phase was interrupted by four intermediate test sessions, which were introduced at regular intervals. This was done to observe the evolution of the pre-existing coordination patterns during the learning of the new pattern. Third, the experiment was conducted over three days, and the number of practice trials during the learning session was increased to 50 (10 the first day, and 20 the second and the third day). Finally, a retention test completed one week after the end of practice was used to estimate the durability of the changes observed in spontaneous and learned patterns.

The two experiments revealed several important results.

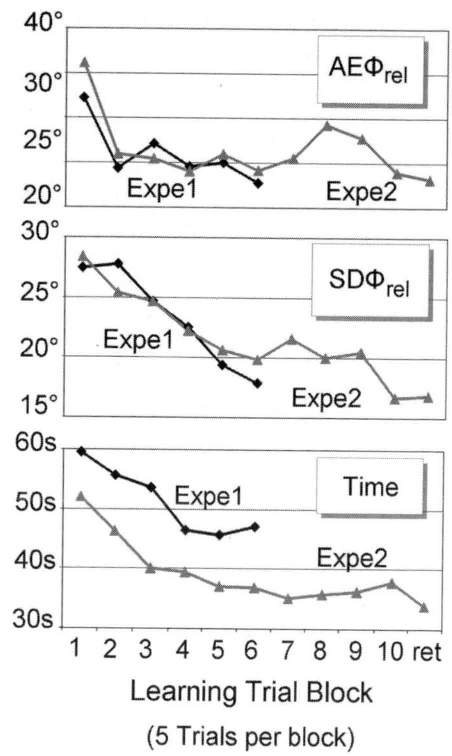

FIGURE 6. Mean results for the learning session in the two experiments (Expe 1 and Expe2). Evolution of absolute error *AE* for relative phase ϕ_{rel}, standard deviation of relative phase $SD\phi_{rel}$, and movement time during learning. Ret indicates the retention test.

Learning Session. We observed a decrease over trial in the absolute error, *AE* (discrepancy between produced ϕ_{rel} and learned ϕ_{rel}), a decrease in $SD\ \phi_{rel}$, and a decrease in the time taken to perform a trial (i.e., movement time *MT*). All changes were significant. Figure 6 presents the evolution of these three variables for the two experiments. The observed evolution of accuracy, stability, and movement time confirms that participants learned the requested 135° coordination with practice.

Influence of Initial Stability of Spontaneous Patterns on Learning. No significant correlation was found between $SD\ \phi_{rel}$ of the initial spontaneous pattern and the three variables capturing learning (*AE*, $SD\ \phi_{rel}$, or *MT*), neither at the beginning nor at the end of the learning session. Thus, contrary to what has been occasionally found in bimanual studies (Zanone & Kelso, 1992), we did not find any relation between initial stability and learning rate.

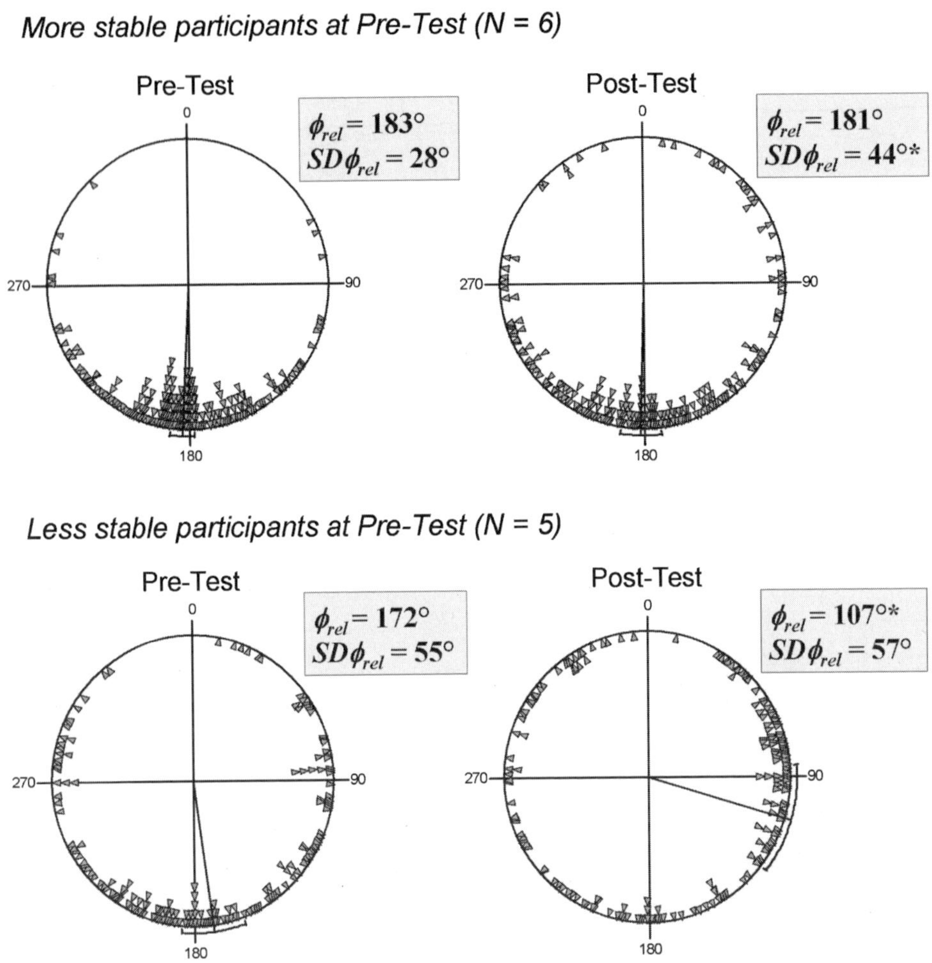

FIGURE 7. Polar distributions of relative phase values at pre-test (left) and post-test (right) in Experiment 1, showing differences in pattern destabilization due to learning. (Top): Values of the most stable participants showing a significant change(*) in standard deviation and confidence interval around the mean relative phase between pre- and post-test; (Bottom) Values of less stable participants showing a significant change in relative phase values between pre- and post-test.

Differential Influence of Learning a New Coordination on Spontaneous Postural Modes. Between the pre-test and the post-test (and between the pre-test and the intermediate tests in Experiment 2), we observed important changes in ϕ_{rel} as well as in $SD\,\phi_{rel}$, providing evidence for the destabilization of initial spontaneous patterns due to learning. However, these changes did not occur equally across all participants and appeared to be dependant upon initial stability (i.e., $SD\,\phi_{rel}$ at pre-test). Indeed, in Experiment 1, participants with high initial stability ($N = 6$) presented a loss of this stability between the pre-test and the post-test, whereas participants with low initial stability ($N = 5$) showed a shift in relative phase toward the learned pattern. In other words, participants modified either the stability of their spontaneous coordination, or the ankle-hip coordination itself, depending on the stability of the spontaneous coordination (Figure 7). The difference in the nature of the destabilization was also observed for the anti-phase pattern of Experiment 2 (i.e., high frequency group). Participants from the high frequency group (0.65 Hz) presented three different types of destabilization (Figure 8): the most stable participants ($N = 2$) did not show any destabilization

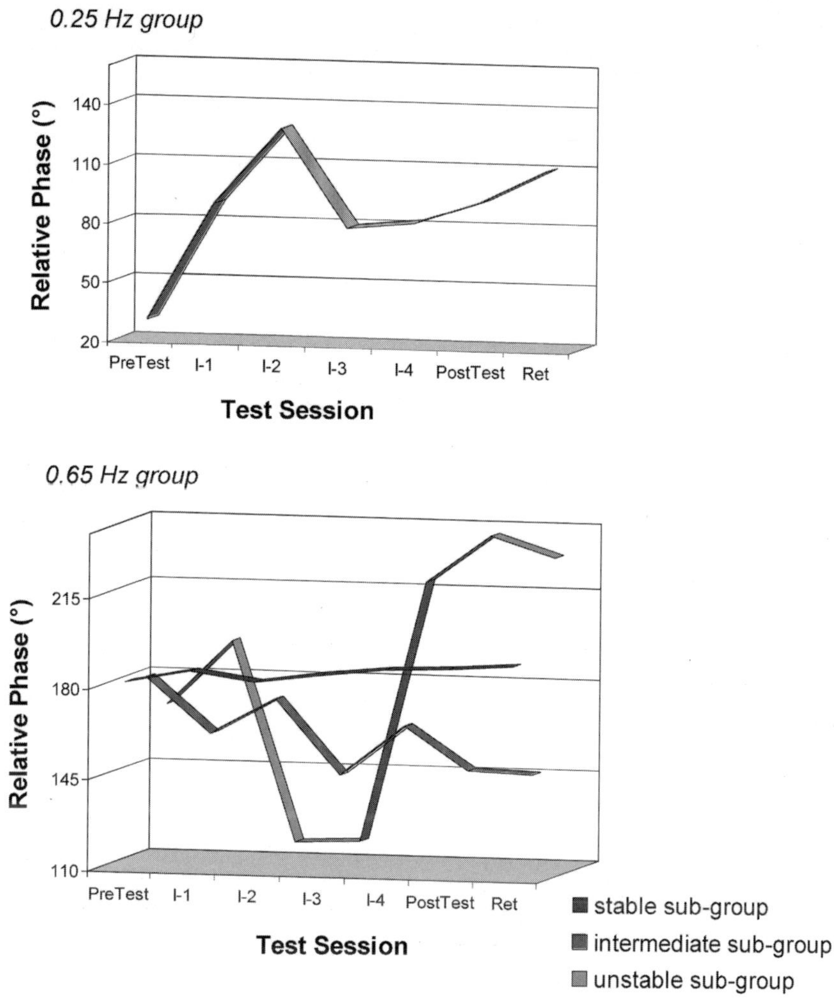

FIGURE 8. Destabilization due to learning in Experiment 2. Changes in the relative phase ϕ_{rel} over the seven tracking tests (Pre-test, I-1 to I-4: inter-test 1 to inter-test 4, Post-test, Ret: retention test), for the 0.25 Hz group (Top) and the three sub-groups of the 0.65 Hz group (bottom).

of the initial spontaneous pattern ($\phi_{rel} = 183°$ at the pre-test and 191° at the post-test); the two intermediately stable participants at pre-test showed a shift in the spontaneous relative phase toward the learned pattern ($\phi_{rel} = 182°$ at the pre-test and 148° at the post-test); the less stable participants ($N = 2$) exhibited the same shift toward the learned pattern, before a clear shift toward its symmetric pattern, i.e., 225° ($\phi_{rel} = 168°$ at pre-test and 236° at post-test). All participants from the low frequency group (Figure 8) evidenced a shift in the coordination ($\phi_{rel} = 31°$ at pre-test) in the direction of the learned pattern ($\phi_{rel} = 93°$ at post-test). As we can see in Figure 8, the changes recorded for the low and high frequency groups between pre- and post-tests were again observed at the retention test, suggesting that these changes were relatively permanent.

Conclusion

To some extent, the present results echo findings from research on bimanual coordination (e.g., see Faugloire, 2005 for a recent review). First, we found that it was possible to learn a new pattern of relative phase, and that learning was fast. Repeating the new coordination pattern produced an increase in accuracy and a decrease in variability (e.g., Fontaine et al., 1997; Lee et al., 1995; Wenderoth & Bock, 2001; Zanone, & Kelso, 1992). The fact that learning was similar across very different systems (the bimanual system, the postural system) reinforces the idea of motor equivalency, the idea that the acquisition of new coordination patterns follows a set of general laws (Bernstein, 1967, Newell, 1996). Second, learning a new (self-paced) 135° coordinative mode was associated with destabilization of postural coordination modes that were assembled in support of the externally-paced tracking task. This destabilization was fast (after 30 learning trials in Experiment 1 and only 10 trials in Experiment 2), suggesting that in learning, stabilization and destabilization are intertwined phenomena, which can occur simultaneously rather than successively. It was also durable, as shown by the retention test.

The present results, together with the modeling effort reported in this chapter, support the conjecture that in-phase and anti-phase patterns observed during standing are emergent properties of the interaction between the natural tendencies of the postural systems and the external and internal constraints that shape coordination dynamics (e.g., Bardy, 2004, Faugloire, Bardy, Merhi & Stoffregen, 2005). The presence of appropriate constraints—not only including environmental or individual constraints but also task goals, such as the instruction to learn a novel pattern—is a prerequisite for the emergence of specific coordination patterns. Some evidence has been found in favor of the idea that, in coordination dynamics, symmetry is preserved across the learning process: at least two participants (see Figure 8) exhibited a shift toward a coordination pattern (225°) that was symmetrical (with respect to the 180° relative phase pattern produced initially) to the pattern that they were attempting to learn (135°). Similar effects have been found in the context of bi-manual coordination (e.g., Zanone & Kelso, 1997), suggesting that for these two participants (at least), the underlying symmetry of the attractor landscape was preserved during postural learning. Such an effect would confirm the abstract, and therefore transferable, nature of learning. The extent to which these patterns of coordination are abstract and transferable will be addressed in future research, with specific transfer experiments between postural and bimanual coordination.

References

Balasubramaniam, R., Riley, M. A., & Turvey, M.T. (2000). Specificity of postural sway to the demands of a precision task. *Gait & Posture, 11,* 12–24.

Bardy, B. G. (2004). Postural coordination dynamics in standing humans. In V. K. Jirsa & J. A. S. Kelso (Eds.), *Coordination dynamics: Issues and trends* (pp. 103–121). Berlin: Springer.

Bardy, B. G., Faugloire, E., & Stoffregen, T. (2005). The dynamics of learning new postural patterns. Manuscript submitted for publication.

Bardy, B. G., Marin, L., Stoffregen, T. A., & Bootsma, R. J. (1999). Postural coordination modes considered as emergent phenomena. *Journal of Experimental Psychology: Human Perception and Performance, 25,* 1284–1301.

Bardy, B. G., Oullier, O., Bootsma, R. J., & Stoffregen, T. A. (2002). Dynamics of human postural transitions. *Journal of Experimental Psychology: Human Perception and Performance, 28,* 499–514.

Barin, K. (1989). Evaluation of a generalized model of human postural dynamics and control in the sagittal plane. *Biological Cybernetics, 61,* 37–50.

Beek, P. J., Peper, C.E., Daffertshofer, A., Van Soest, A., & Meijer, O. G. (1998). Studying perceptual-motor actions from mutually constraining perspectives. In: A. A. Post, J. R. Pijpers, P. Bosch, M. J. S. Boschker (eds.), *Models in Human Movement Sciences*: Proceedings of the second symposium of the Institute for Fundamental and Clinical movement Science (pp. 93–111). Enschede, NL: Print-Partners Ipskamp.

Bernstein, N. A. (1967). *The co-ordination and regulation of movements.* Oxford: Pergamon Press.

Dijkstra, T. M. H., Schöner, G., & Gielen, C. C. A. M. (1994). Temporal stability of the action-perception cycle for postural control in a moving visual environment. *Experimental Brain Research, 97,* 477–486.

Farley, C. T., & Morgenroth, D. C. (1999). Leg stiffness primarily depends on ankle stiffness during human hopping. *Journal of Biomechanics, 32,* 267–273.

Faugloire, E. (2005). *Approche dynamique de l'apprentissage des coordinations posturales* [Dynamical perspective on learning postural coordination modes]. PhD thesis in Movement Sciences, University of Paris 11.

Faugloire, E., Bardy, B. G., Merhi, O., & Stoffregen, T.A (2005). Exploring coordination dynamics of the postural system with real-time visual feedback. *Neuroscience Letters, 374,* 136–141.

Fontaine, R. J., Lee, T. D., & Swinnen, S. P. (1997). Learning a new bimanual coordination pattern: reciprocal influences of intrinsic and to-be-learned patterns. *Canadian Journal of Experimental Psychology, 51,* 1–9.

Fourcade, P., Bardy, B. G., & Roudeix, S. (2005). A 'compound' model of human postural transitions. Manuscript in preparation.

Haken, H., Kelso, J. A. S., & Bunz, H. (1985). A theoretical model of phase transitions in human hand movements. *Biological Cybernetics, 51,* 347–356.

Kay, B.A., & Warren, W.H. (2001). Coupling of posture and gait: mode locking and parametric excitation. *Biological Cybernetics, 85,* 89–106.

Kelso, J. A. S. & Zanone, P. G. (2002). Coordination dynamics of learning and transfer across different effector systems. *Journal of Experimental Psychology: Human Perception and Performance, 28,* 776–797.

Kelso, J. A. S. (1984). Phase transitions and critical behavior in human bimanual coordination. *American Journal of Physiology: Regulatory, Integrative, and Comparative, 15,* R1000–R1004.

Kuo, A. D. (1995). An optimal control model for analyzing human postural balance. *IEEE Transaction Biomed Engineering, 42,* 87–101.

Le Veau, B. (1977). *Biomechanics of human motion.* Philadelphia, PA: W. B. Saunders Company.

Lee, T. D., Swinnen, S. P., & Verschueren, S. (1995). Relative phase alterations during bimanual skill acquisition. *Journal of Motor Behavior, 27,* 263–274.

Marin, L., Bardy, B. G., Baumberger, B., Flückiger, M., & Stoffregen, T. A. (1999). Interaction between task demands and surface properties in the control of goal-oriented stance. *Human Movement Science, 18,* 31–47.

Nashner, L. M, & McCollum, G. (1985). The organization of postural movements: A formal basis and experimental synthesis. *Behavioral and Brain Sciences, 8,* 135–172.

Nashner, L.M. (1976). Adapting reflexes controlling the human posture. *Experimental Brain Research, 26,* 59–72.

Newell, K. M. (1986). Constraints on the development of coordination. In M. G. Wade & H. T. A. Whiting (Eds.), *Motor development in children: Aspects of coordination and control* (pp. 341–360). Dordrecht: Martinus Nijhoff.

Newell, K. M. (1996). Change in movement and skill: Learning, retention, and transfer. In M. L. Latash & M. T. Turvey (Eds.), *Dexterity and its development* (pp. 393–430). Mahwah, NJ: L. Erlbaum Associates.

Oullier, O., Bardy, B. G., Stoffregen, T. A., & Bootsma, R. J. (2002). Postural coordination in looking and tracking tasks. *Human Movement Science, 21,* 147–167.

Oullier, O., Bardy, B. G., Stoffregen, T. A., & Bootsma, R. J. (2004). Task-specific stabilization of postural coordination during stance on a beam. *Motor Control, 7,* 174–187.

Schöner, G. (1991). Dynamic theory of action-perception patterns: The "moving room" paradigm. *Biological Cybernetics, 64,* 455–462.

Smethurst, C. J. & Carson, R. G. (2001). The acquisition of movement skills: Practice enhances the dynamic stability of bimanual coordination. *Human Movement Science, 20,* 499–529.

Stefanyshyn, D. J., & Nigg, B.M. (1998). Dynamic angular stiffness of the ankle joint during running and sprinting. *Journal of Applied Biomechanics, 14,* 292–299.

Sternad, D., Amazeen, E. L., & Turvey, M. T. (1996). Diffusive, synaptic, and synergetic coupling: An evaluation through in-phase and antiphase rhythmic movements. *Journal of Motor Behavior, 28,* 255–269.

Stockwell, C. W., Koozekanani, S. H., & Barin, K. (1981). A physical model of human postural dynamics. *Annals New York Academy of Sciences, 374,* 722–730.

Stoffregen, T. A., Smart, L. J., Bardy, B. G., & Pagulayan, R. J. (1999). Postural stabilization of looking. *Journal of Experimental Psychology: Human Perception and Performance, 25,* 1641–1658.

Taga, G. (1994). Emergence of bipedal locomotion through entrainment among the neuro-musculo-skeletal system and the environment. *Physica D, 75,* 190–208.

Taga, G. (1995). A model of the neuro- musculo-skeletal system for human locomotion. I. Emergence of basic gait. *Biological Cybernetics, 73,* 97–111.

Weiss, P. L., Hunter, I.W., & Kearney, R.E. (1988). Human ankle joint stiffness over the full range of muscle activation levels, *Journal of Biomechanics, 21,* 539–544.

Wenderoth, N. & Bock, O. (2001). Learning of a new bimanual coordination pattern is governed by three distinct processes. *Motor Control, 1,* 23–35.

Winter, D. A. (1990). *Biomechanics and motor control of human gait.* Waterloo: University of Waterloo Press.

Yang, J. F., Winter, D. A., Wells, R.P. (1990). Postural dynamics in the standing human. *Biological Cybernetics, 62,* 309–320.

Zanone, P. G. & Kelso, J. A. S. (1992). Evolution of behavioral attractors with learning: Nonequilibrium phase transitions. *Journal of Experimental Psychology: Human Perception and Performance, 18,* 403–421.

Zanone, P. G. & Kelso, J. A. S. (1997). Coordination dynamics of learning and transfer: Collective and component levels. *Journal of Experimental Psychology: Human Perception and Performance, 23,* 1454–1480.

8. THE ROLE OF THE MOTOR CORTEX IN MOTOR LEARNING

Mark Hallett

Human Motor Control Section, NINDS, NIH, Bethesda, MD USA

Abstract

The motor cortex is clearly more than a simple executor of motor commands and is likely involved with different aspects of motor learning. The motor cortex shows considerable plasticity, and both excitability and amount of territory devoted to a muscle or to a specific task can expand or shrink depending on the amount of use. There are also short-term increases in motor cortex activity when learning new tasks. In the serial reaction time task (SRTT), as demonstrated by transcranial magnetic stimulation (TMS), EEG, and positron emission tomography, the motor cortex is involved in the implicit phase of motor learning and declines in activity during the explicit phase. In learning to increase pinch force and pinch acceleration between index and thumb, the motor evoked potential (MEP) from TMS increases during the early stage of learning, but then declines even though the behavioral change is maintained. In learning a bimanual task, there is a transient increase in EEG coherence between the two hemispheres at the time of the learning. What function this short-term increase in motor cortex activity serves is not certain. It has recently been established that motor learning goes through a phase of consolidation and becomes more secure simply with the passage of time. This was first demonstrated while adapting to making accurate movements in a force field. Neuroimaging studies with these same movements in a force field show a transient increase in motor cortex activity during the learning phase. In our laboratory, we have studied consolidation of the learning to increase pinch force and acceleration. Consolidation is disrupted by 1 Hz repetitive TMS of the motor cortex if done immediately after learning, but not after a rest of 6 hours. This demonstrates a role of the motor cortex in consolidation.

The motor cortex is the primary control center of the human brain for control of movement. It contributes a large percentage of the axons to the corticospinal tract and virtually all of the axons that are monosynaptic onto alpha-motoneurons. The most obvious deficit after lesions of the motor cortex is paresis. It is now clear, however, that the motor cortex is not just an executor of movement. One of its other roles is to contribute to motor learning. Motor learning is a type of procedural learning, and can be defined as a change in motor performance with practice.

In the past decade, considerable evidence has accumulated about the plasticity of the human motor cortex as a function of use and motor learning. Using TMS, it is possible to map the degree and extent of excitability of individual muscles on the scalp surface. Body parts that are used more have a larger representation. This was first demonstrated by looking at the cortical representation area of the first dorsal interosseous muscle (FDI) of blind individuals that read Braille many hours per day (Pascual-Leone et al. 1993a). The FDI moves the index finger over the Braille characters. The representation of the FDI was enlarged in the hemisphere opposite the reading hand, but not in the other hemiphere nor in either hemiphere of blind individuals that did not read Braille. Conversely, representations will shrink if the body part is not used. This was first demonstrated by looking at the cortical representation of the tibialis anterior muscle in individuals who had their ankles immobilized in a cast following an ankle injury (Liepert et al. 1995). The representation was smaller (or, as least, less excitable) following the immobilization.

The representation of a body part increases when it is involved in a motor learning task. We mapped the cortical motor areas targeting the forearm finger flexor and extensor muscles in normal subjects learning a one-handed, five-finger exercise on an electronic piano (Pascual-Leone et al. 1995). The task was metronome-paced so that improvement in accuracy should identify skill learning. The piano was connected by a MIDI interface to a personal computer for quantification of times of key presses. Subjects practiced the task for 2 hours daily. They improved in terms of ability to keep accurate time with the metronome and in reduction of errors. The size of the representation expanded over 5 days as the task was learned.

From basic principles, it is reasonable to consider the motor cortex as a relevant site for motor skill learning. Cortical cells have complex patterns of connectivity including variable influences on multiple muscles within a body part. Changes in these patterns could give rise to alterations in representation areas. Such changes could occur quickly with alterations in firing patterns of inhibitory interneurons. Long-term potentiation (LTP) could lead to more long lasting change, and this phenomenon has been demonstrated in the motor cortex (Iriki et al. 1989).

Further evidence that cortical map plasticity is important in skill learning comes from primate experiments. Lesions of the basal forebrain cholinergeric pathways (that blocked cortical map plasticity) inhibited skill learning (Conner et al. 2003). This lesion, however, did not block associative fear learning, indicating differences in different types of learning.

Using neuroimaging with motor learning tasks, it is clear that a variety of brain regions are involved depending on the task. (Friston et al. 1992; Grafton et al. 1992; Seitz and Roland 1992; Grafton et al. 1994; Jenkins et al. 1994; Schlaug et al. 1994; Seitz et al. 1994; Karni et al. 1995) The primary motor cortex has almost always been activated to some extent although because of resolution it has often been difficult to separate primary motor cortex from premotor cortex and/or primary sensory cortex. Moreover, the results have been somewhat confusing because techniques and experimental paradigms have differed, and because motor performance was not necessarily held constant over the course of learning.

One well known study used fMRI and focused attention on the contralateral primary motor cortex (Karni et al. 1995). Two finger tapping sequences were compared, one that was in the process of being learned and a second that was already learned. Although the learned sequence could have been performed faster, both sequences were performed at the same rate paced by an auditory stimulus. As the motor task was learned, more area of the motor cortex was activated.

In most of these studies, cerebellar activation is also evident in the learning phase and declines when the movement is learned. This certainly indicates a role in learning. That the cerebellar activation declines when the movement is learned is against the ideas that the cerebellum stores the movement and is in some way responsible for the automatic running of the motor program when it is well learned. Late in motor skill learning, another relatively common neuroimaging result is that there is activation of parietal and premotor areas (Grafton et al. 1994; Jenkins et al. 1994; Seitz et al. 1994). Sometimes the basal ganglia, particularly the putamen, are also activated (Grafton et al. 1994).

Many of the studies of motor learning are complicated, and it is difficult to separate out the different facets. One facet is learning the order of a number of components of a complex movement with sequential elements. The SRTT appears to be a nice paradigm to study motor learning of sequences. The ability to carry out sequences of motor actions is clearly a critical part of most complex tasks, and the SRTT should be able to help understand this aspect of learning. The task is a choice reaction time with typically four possible responses. The responses can be carried out by key presses with four different fingers. A visual stimulus indicates which is the appropriate response. The completion of one response triggers the next stimulus. Each movement is simple and separate from the others so that the movement aspect of this task is different (and easier) than other tasks considered previously such as finger tapping or piano playing. The trick in this task is that unbeknownst to the naive subject the stimuli are a repeating sequence. With practice at this task, the responses get faster even though the subject has no conscious recognition that the sequence is repetitive. This is called implicit learning. With continuing practice and improvement, there is recognition that there is a sequence, but it may not be possible to specify what it is. Now knowledge is becoming explicit. With even more practice, the sequence can be specified and it has become declarative as well as procedural. Performance gets even better at this stage, but the subject's strategy can change since the stimuli can be anticipated.

Thus, the SRTT appears to assess two processes relating to the sequencing of motor behavior while factoring out elements of motor coordination. As such, it might be considered a test of some components of motor skill learning.

We have looked at the intermanual transfer of implicit learning of the SRTT (Wachs et al. 1994). After

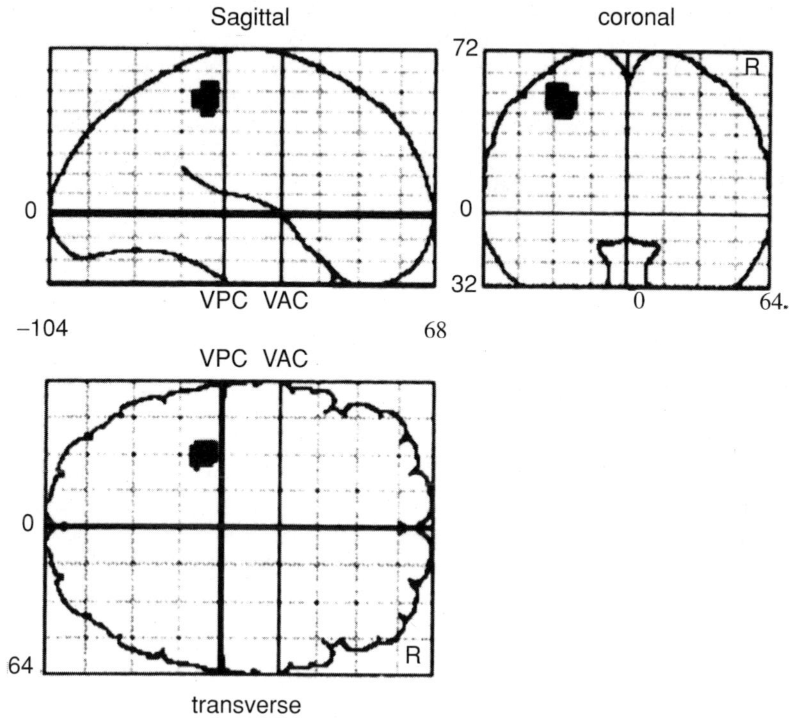

FIGURE 1. PET study of serial reaction time task showing that the motor cortex is active with implicit learning. Illustrated are sites of activation that correlate with reduction of response time in scans during blocks where there was no explicit knowledge of the sequence. From Honda et al. (1998) with permission.

a few blocks of training with one hand, subsequent blocks were done with the other hand. Four groups of normal subjects were studied each with one condition: (1) random sequence, (2) a new sequence, (3) parallel image of the original sequence, and (4) mirror image of the original sequence. Only group 4 showed a carry-over effect from the original learning. This result suggests that what is stored as implicit learning is a specific sequence of motor outputs and not a spatial pattern.

Implicit learning in the SRTT is impaired in patients with cerebellar degeneration, Parkinson's disease, Huntington's disease, and progressive supranuclear palsy (Pascual-Leone et al. 1993b) Patients with cerebellar degeneration were particularly severely affected. Not only was performance characterized by lack of improvement in reaction time, there was also lack of development of explicit knowledge. Moreover, even giving the patients information about the sequence in advance (explicit knowledge), did not help improve reaction time. On the other hand, implicit learning is preserved in patients with temporal lobe lesions and patients with short-term declarative memory disturbances such as most patients with Alzheimer's disease.

In relation to the question of the involvement of the primary motor cortex in implicit learning, we mapped the motor cortex with TMS contralateral to the hands of normal subjects performing the SRTT (Pascual-Leone et al. 1994). Mapping was done at intervals while the subjects were at rest between blocks of the SRTT. The map gradually enlarged during the implicit and explicit learning phases, but as soon as full explicit learning was achieved, the map size returned to baseline. This suggests an important role for primary motor cortex in this task.

We examined the dynamic involvement of different brain regions in implicit and explicit motor sequence learning using PET (Honda et al. 1998). In an SRTT, subjects pressed each of four buttons with a different finger of the right hand in response to a visually presented number. Test sessions consisted of

10 cycles of the same 10-item sequence. The effects of explicit and implicit learning were assessed separately using a different behavioral parameter for each type of learning: correct recall of the test sequence for explicit learning and improvement of reaction time before the successful recall of any component of the test sequence for implicit learning. During the implicit learning phase, when the subjects were not aware of the sequence, improved reaction time was associated with increased activity in the contralateral primary sensorimotor cortex (Fig. 1). Explicit learning, shown as a positive correlation with the correct recall of the sequence, was associated with increased activity in the posterior parietal cortex, precuneus and premotor cortex bilaterally, also in the supplementary motor area predominantly in the left anterior part, left thalamus, and right dorsolateral prefrontal cortex.

There have been a large number of other neuroimaging studies of the SRTT. Grafton et al. (Grafton et al. 1995) studied two situations. In one, there was a second, distracting task to be done at the same time as the SRTT. Such distraction does not interfere with implicit learning, but makes explicit learning much less likely. Hence, regions that were active are likely to reflect implicit learning. In a second experiment, there was no other task, and subjects were scanned in the explicit learning phase. In the implicit learning situation, there was activation of the contralateral primary motor cortex, SMA and putamen. In the explicit learning situation, there was activation of the ipsilateral DLPFC and premotor cortex and of the parietal cortex bilaterally. Doyon et al. (Doyon et al. 2002; Doyon et al. 2003) emphasized early cerebellar activation, a middle stage with premotor, anterior cingulated and parietal activation, and a later stage with putamen, SMA, precuneus and prefrontal activation. Penhune et al. (Penhune and Doyon 2002) investigated an SRTT with a different type of sequence; there was only one key, but the elements were of different duration. This begins to get at the issue of rhythm. Here again the cerebellum was active early and later in learning, the activation shifted to basal ganglia and medial frontal areas. Several days later, imaging during recall showed activation of primary motor cortex, premotor cortex, and parietal cortex, but not cerebellum or basal ganglia. Seidler et al. (Seidler et al. 2002) using an experimental paradigm similar to that of Grafton et al. showed specifically that the cerebellum is not involved in early implicit learning in the SRTT. Using a distractor task during the SRTT, there was no cerebellar activation, but evaluation afterwards showed that implicit learning had indeed occurred. Cerebellar activation was present, however, upon first demonstration of the implicit learning after the distractor task was discontinued. The implication was made that the cerebellar contribution related more to performance than learning itself.

Added evidence for the role of the motor cortex in SRTT learning comes from a study of transcranial direct current stimulation (TDC). Anodal TDC, that enhances cortical excitability, improves implicit learning in the SRTT while similar stimulation of premotor and prefrontal stimulation does not (Nitsche et al. 2003).

To summarize the studies of the SRTT, it appears that multiple structures in the brain are involved, and that involvement comes at different stages. The primary motor cortex appears to play a definite role in implicit learning. Premotor and parietal cortical areas appear to play a role in explicit learning, perhaps in part by storage of the sequence. This concept is supported by the clinical finding that damage of premotor and parietal areas can lead to apraxia; this might be interpreted as a deficiency of motor memories for complex movements. The cerebellum also appears relevant in learning movement sequences given the results in patients with cerebellar degeneration, but the nature of the role may relate more to the ability to manifest what is learned. The basal ganglia role is more obscure.

In addition to a role in implicit learning, the motor cortex may also contribute to the process of consolidation. Consolidation is the process whereby learned skills become more permanent. Immediately after learning, the motor memory is fragile. In particular, it is vulnerable to disruption by learning of something similar. However, if there is no disruption, with the passage of time, the memory becomes more robust. It is this process, of becoming more robust with time, that is designated consolidation. Consolidation was demonstrated for the first time clearly in the motor system with the study of Brashers-Krug et al. (Brashers-Krug et al. 1996). These investigators studied subjects making center-out movements on a two-dimensional surface under the influence of various force fields. Without the force field, the movements are made in straight paths. When first experiencing the field, the movements become distorted, but with practice, the movements can become straight even in the force field. If a force field is learned, then the performance on the field is maintained the next day. If a different force field is learned immediately after the first, the learning of the first field is completely lost. This disruption by a second force field does not occur; however, if there is the passage of 4 to 6 hours between learning of the two fields. This demonstrates that consolidation of learning of the first field occurs during this several hour period.

Imaging studies have been done with force field learning, and early in learning, there was activation of motor cortex, putamen and prefrontal cortex (Shadmehr and Holcomb 1997). In the recall of the force field, activation was now primarily in parietal and premotor cortex and cerebellum. The pattern of early learning and late recall is similar to the pattern seen by Honda et al. with SRTT learning. While the authors of this study interpreted the early activation of the motor cortex to be due to a longer movement trajectory that occurred in the early phases of learning, it seems more likely that it was engaged because of implicit learning.

Force field learning is a nice model and has been used to advantage to illustrate certain principles. It is a complex task, however, and while often referred to as an example of adaptation learning, it is likely a combination of adaptation and skill learning.

We tested the possibility that the human M1 is essential to early motor consolidation (Muellbacher et al. 2002). We monitored changes in elementary motor behavior of pinching between the thumb and index finger while subjects practiced fast finger movements that rapidly improved in movement acceleration and muscle force generation. Low-frequency, repetitive TMS of M1 but not frontal or occipital cortex specifically disrupted the retention of the behavioral improvement, but did not affect basal motor behavior, task performance, or motor learning by subsequent practice (Fig. 2). However, if the repetitive TMS was given 6 hours after practice, then it no longer disrupted the recall of the newly acquired motor skill (Fig. 3). These findings indicate that the human M1 is specifically engaged during the early stage of motor consolidation.

Motor learning is a complex phenomenon with many components. Depending on the particular task, different anatomical structures are involved. It would be an oversimplification to say that only one part of the brain is involved with any task; it is more likely that a network is functional. On the other hand, it is possible to identify some aspects where particular structures play a major role. The development of new skills has many facets and likely engages large portions of the brain. The motor cortex is involved early, plays a role in implicit learning and consolidation and, by map plasticity may assign resources to different movements.

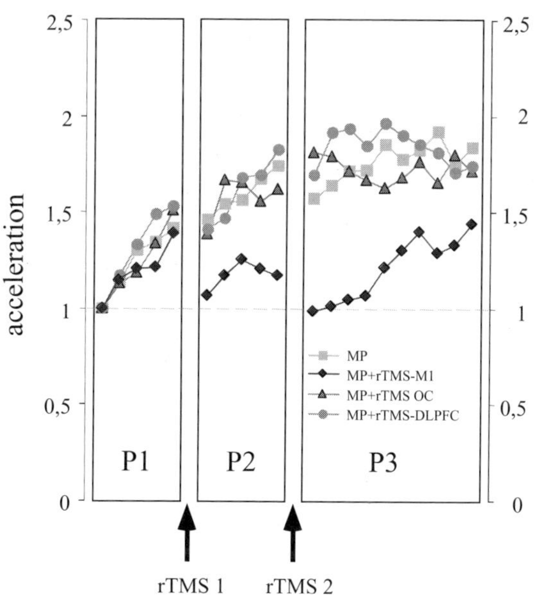

FIGURE 2. Acceleration of pinching force with practice and various interventions. P1, P2 and P3 are practice periods. Repetitive TMS is given between the practice periods. Stimulation over M1, but not occipital cortex (OC) or dorsolateral prefrontal cortex (DLPFC), blocked the consolidation of the learning. From Muellbacher et al. (2002) with permission.

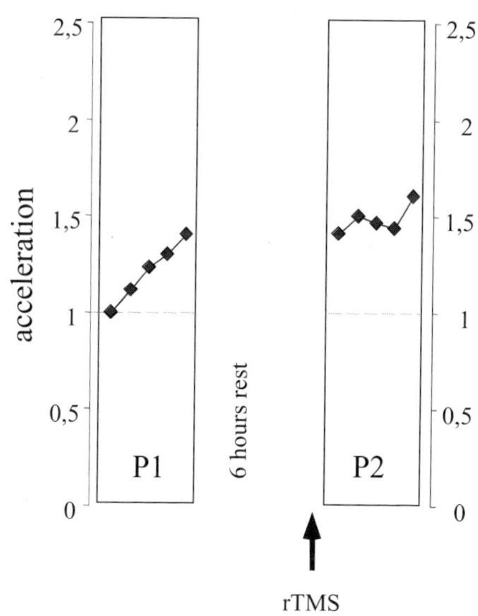

FIGURE 3. Acceleration of pinching force in two practice periods, and with 6 hours rest and then repetitive TMS of the primary motor cortex between the periods. P1 and P2 are practice periods. Stimulation over M1 in this circumstance does not block the consolidation of learning. From Muellbacher et al. (2002) with permission.

Acknowledgement

This chapter is revised and excerpted from another recent chapter (Hallett 2004) which is itself updated from a previous chapter (Hallett and Grafman 1997). Work of the US government has no copyright.

References

Brashers-Krug T, Shadmehr R, Bizzi E (1996) Consolidation in human motor memory. Nature 382: 252–255.

Conner JM, Culberson A, Packowski C, Chiba AA, Tuszynski MH (2003) Lesions of the Basal forebrain cholinergic system impair task acquisition and abolish cortical plasticity associated with motor skill learning. Neuron 38: 819–829.

Doyon J, Penhune V, Ungerleider LG (2003) Distinct contribution of the cortico-striatal and cortico-cerebellar systems to motor skill learning. Neuropsychologia 41: 252–262.

Doyon J, Song AW, Karni A, Lalonde F, Adams MM, Ungerleider LG (2002) Experience-dependent changes in cerebellar contributions to motor sequence learning. Proc Natl Acad Sci U S A 99: 1017–1022.

Friston KJ, Frith CD, Passingham RE, Liddle PF, Frackowiak RSJ (1992) Motor practice and neurophysiological adaptation in the cerebellum: a positron tomography study. Proc Royal Soc London (Biology) 248: 223–228.

Grafton ST, Hazeltine E, Ivry R (1995) Functional mapping of sequence learning in normal humans. J Cogn Neurosci 7: 497–510.

Grafton ST, Mazziotta JC, Presty S, Friston KJ, Frackowiak RSJ, Phelps ME (1992) Functional anatomy of human procedural learning determined with regional cerebral blood flow and PET. J Neurosci 12: 2542–2548.

Grafton ST, Woods RP, Tyszka M (1994) Functional imaging of procedural motor learning: Relating cerebral blood flow with individual subject performance. Human Brain Mapp 1: 221–234.

Hallett M (2004) Motor learning. In: Freund HJ, Jeannerod M, Hallett M, Leiguarda R (eds) Higher Order Motor Disorders. Oxford University Press, Oxford, in press.

Hallett M, Grafman J (1997) Executive function and motor skill learning. In: Schmahmann JD (ed) The Cerebellum and Cognition, vol 41. Academic Press, San Diego, pp 297–323.

Honda M, Deiber MP, Ibanez V, Pascual-Leone A, Zhuang P, Hallett M (1998) Dynamic cortical involvement in implicit and explicit motor sequence learning. A PET study. Brain 121: 2159–2173.

Iriki A, Pavlides C, Keller A, Asanuma H (1989) Long-term potentiation of motor cortex. Science 245: 1385–1387.

Jenkins IH, Brooks DJ, Nixon PD, Frackowiak RSJ, Passingham RE (1994) Motor sequence learning: a study with positron emission tomography. J Neurosci 14: 3775–3790.

Karni A, Meyer G, Jezzard P, Adams M, Turner R, Ungerleider LG (1995) Functional MRI evidence for adult motor cortex plasticity during motor skill learning. Nature 377: 155–158.

Liepert J, Tegenthoff M, Malin JP (1995) Changes of cortical motor area size during immobilization. Electroenceph Clin Neurophysiol 97: 382–386.

Muellbacher W, Ziemann U, Wissel J, Dang N, Kofler M, Facchini S, Boroojerdi B, Poewe W, Hallett M (2002) Early consolidation in human primary motor cortex. Nature 415: 640–644.

Nitsche MA, Schauenburg A, Lang N, Liebetanz D, Exner C, Paulus W, Tergau F (2003) Facilitation of implicit motor learning by weak transcranial direct current stimulation of the primary motor cortex in the human. J Cogn Neurosci 15: 619–626.

Pascual-Leone A, Cammarota A, Wassermann EM, Brasil-Neto JP, Cohen LG, Hallett M (1993a) Modulation of motor cortical outputs to the reading hand of Braille readers. Ann Neurol 34: 33–37.

Pascual-Leone A, Dang N, Cohen LG, Brasil-Neto JP, Cammarota A, Hallett M (1995) Modulation of muscle responses evoked by transcranial magnetic stimulation during the acquisition of new fine motor skills. J Neurophysiol 74: 1037–1045.

Pascual-Leone A, Grafman J, Clark K, Stewart M, Massaquoi S, Lou J-S, Hallett M (1993b) Procedural learning in Parkinson's disease and cerebellar degeneration. Ann Neurol 34: 594–602.

Pascual-Leone A, Grafman J, Hallett M (1994) Modulation of cortical motor output maps during development of implicit and explicit knowledge. Science 263: 1287–1289.

Penhune VB, Doyon J (2002) Dynamic cortical and subcortical networks in learning and delayed recall of timed motor sequences. J Neurosci 22: 1397–1406.

Schlaug G, Knorr U, Seitz RJ (1994) Inter-subject variability of cerebral activations in acquiring a motor skill: a study with positron emission tomography. Exp Brain Res 98: 523–534.

Seidler RD, Purushotham A, Kim SG, Ugurbil K, Willingham D, Ashe J (2002) Cerebellum activation associated with performance change but not motor learning. Science 296: 2043–2046.

Seitz RJ, Canavan AG, Yaguez L, Herzog H, Tellmann L, Knorr U, Huang Y, Homberg V (1994) Successive roles of the cerebellum and premotor cortices in trajectorial learning. NeuroReport 5: 2541–2544.

Seitz RJ, Roland PE (1992) Learning of sequential finger movements in man: a combined kinematic and positron emission tomography (PET) study. Eur J Neurosci 4: 154–165.

Shadmehr R, Holcomb HH (1997) Neural correlates of motor memory consolidation. Science 277: 821–825.

Wachs J, Pascual-Leone A, Grafman J, Hallett M (1994) Intermanual transfer of implicit knowledge of sequential finger movements (abstract). Neurology 44 (Suppl 2): A329.

9. FEEDBACK REMAPPING AND THE CORTICAL CONTROL OF MOVEMENT

Michael S. A. Graziano

Department of Psychology, Princeton University, Princeton NJ 08544

Abstract

Motor cortex in the primate brain controls movement at a complex level. For example, electrical stimulation of motor cortex on a behavioral time scale can elicit multi-joint movements that resemble common gestures in the monkey's behavioral repertoire. How is this complex control accomplished? It was once hypothesized that motor cortex contains a topographic, one-to-one map from points in cortex to muscles. It is now well known that the topography contains a considerable degree of overlap and that the mapping between points in cortex and muscles is many-to-many. However, can a fixed, many-to-many map account for the complex manner in which motor cortex appears to control movement? Recent experiments suggest that the mapping between cortex and muscles may be of a higher order than a fixed, many-to-many map; it may continuously change depending on proprioceptive feedback from the limb. This "feedback remapping" may be a fundamental aspect of motor control, allowing motor cortex to flexibly control almost any high-level or low-level aspect of movement.

Introduction

A central issue in the cortical control of movement is the nature of the map in motor cortex. Neurons in motor cortex map in some fashion to muscles in the periphery, but what are the properties of the map? Is the map one-to-one, in which each location in cortex projects to a single muscle? Is it many-to-many, in which each cortical point connects to many muscles, and each muscle receives input from many cortical locations? Is the map a fixed one, or does it change depending on other sources of input that modulate the pathways between cortex and muscles?

Anatomically, primary motor cortex has a relatively direct, descending projection to the muscles. Pyramidal tract neurons in layer V of cortex project to the spinal cord, where they synapse onto spinal interneurons and in some cases directly onto motoneurons (He et al. 1993; Landgren et al. 1962; Lemon et al. 2004; Maier et al. 2002; Murray & Colter 1981). A range of studies suggest that the neuronal activity in motor cortex is tightly coupled to muscle output. For example, during voluntary movement, the activity of motor cortex neurons is correlated with muscle force and muscle activity (Evarts 1968; Holderfer & Miller 2002; Morrow & Miller 2003). The technique of "spike triggered averaging" shows that an action potential in a neuron in cortex can be followed at short latency by a transient change in muscle activity (Cheney & Fetz 1985; Fetz and Cheney 1980; Lemon et al. 1986; McKiernan et al. 1998). An electrical pulse applied to a point in motor cortex evokes a reliable, short latency effect in a specific set of muscles (Cheney et al. 1985; Maier et al. 1997; Olivier et al. 2001; Park et al. 2001). For these reasons, it appears that motor cortex exerts a relatively direct control over muscles.

The mapping from cortex to muscles, however, is not a punctate, one-to-one map as was once thought (Foerster 1936; Fulton 1938), but instead a many-to-many map (Donoghue et al. 1992; Gould et al. 1986; Jankowska et al. 1975; Kwan et al. 1978; Park et al. 2001; Sanes et al. 1995; Schieber & Hibbard 1993; Schneider et al. 2001). For example, the firing of a single neuron in cortex might be positively correlated with the activity of a set of homonymous muscles and negatively correlated with a set of antagonist muscles

(Cheney & Fetz 1985). This functional linking of a single cortical neuron to many muscles may occur at a variety of levels. It may be partly the result of lateral connections within motor cortex (Baker et al. 1998; Capaday et al. 1998; Gatter et al. 1978; Ghosh & Porter 1988; Huntley & Jones 1991; Kang et al. 1988; Kwan et al. 1987; Landry et al. 1980; Matsumura et al. 1996; Schneider et al. 2002); partly the result of the divergent projection from single neurons in the cortex to multiple target neurons in the spinal cord (Asanuma et al. 1979; Kuang & Kalil 1990; Shinoda et al. 1976); and partly the result of the propriospinal and other interneurons within the spinal cord that link the control of different muscles into functionally useful groups (Bizzi et al. 2000; Jankowska and Hammer 2002; Tantisira et al. 1996). This complexity at every level of the pathway from cortex to muscle results in the many-to-many mapping in which each cortical neuron influences many muscles and each muscle is influenced by many cortical neurons.

One hypothesis is that a fixed, many-to-many mapping from cortex to muscles provides an essentially accurate description of the system, and furthermore can explain how the motor cortex controls movement in such a complex manner. Neurons in motor cortex are active in correlation with a range of movement parameters including direction of movement of the hand through space, velocity, force, joint angle, and arm posture (e.g. Evarts 1968; Caminiti et al. 1990; Georgopoulos et al. 1986; Georgopoulos et al. 1989; Kakei et al. 1999; Kalaska et al. 1989; Reina et al. 2001; Scott & Kalaska 1995; Scott & Kalaska 1997; Sergio & Kalaska 2003). Stimulation of motor cortex on a behavioral time scale can evoke complex, multijoint movements that appear to match the monkey's normal behavioral repertoire (Cooke & Graziano 2004; Graziano et al. 2002a,b; Graziano et al. 2004). Can such complex, higher-order control of movement have as its basis a fixed, many-to-many map from cortex to muscles? One model of cortical function (Todorov 2000) shows that a surprising range of movement parameters can indeed be controlled through a many-to-many muscle map, once the physical properties of the muscles are taken into account.

However, a fixed, many-to-many mapping from cortex to muscles may be an oversimplification. A variety of results suggest that the mapping from cortical neurons to muscles may change from moment to moment, depending on feedback information regarding the kinematic state of the limb (Armstrong & Drew 1985; Bennett & Lemon 1994; Graziano et al. 2004; Kakei et al. 1999; Lemon et al. 1995; Rho et al. 1999; Sanes et al. 1992). Proprioceptive signals from the periphery reach the spinal cord and the cortex, and thus are in a position to modulate the flow of information from neurons in cortex to the muscles. The firing of an output neuron in motor cortex therefore might have very different consequences, resulting in very different patterns of muscle activation, depending on the kinematic state of the limb. In this hypothesis, the mapping from cortical neurons to muscles may not be fixed, but rather may be continuously remapped.

Feedback remapping might allow for a reconciliation between two views of motor cortex. The first view is that there is a direct mapping from the cortical output neurons to the muscles (e.g. Asanuma 1975; Cheney et al. 1985; Evarts 1968; Holderfer & Miller 2002; Lemon et al. 1986). The second view is that motor cortex neurons control high-level movement parameters (Caminiti et al. 1990; Georgopoulos et al. 1986; Georgopoulos et al. 1989; Kakei et al. 1999; Kalaska et al. 1989; Reina et al. 2001). This debate has sometimes been termed the "muscles vs movements" debate. The view of feedback remapping is that there is indeed a mapping from cortex to muscles, but that the mapping is continually adjusted on the basis of kinematic feedback, thereby providing the flexibility to control almost any high-level or low-level aspect of movement.

In this view, feedback remapping is a more fundamental principle than any specific movement coding scheme. Finding the "correct" coding scheme by which motor cortex controls movement, determining whether that scheme is a velocity code, a force code, a direction code, or a postural code, may be misguided, since different tasks might require the control and optimization of different movement parameters (Todorov & Jordan, 2002).

Examples of Feedback Remapping

Sanes et al. (1992) provided one of the first demonstrations of proprioceptive feedback changing the mapping between motor cortex and muscles. They used intracortical microstimulation to map motor cortex in the rat, and found that by placing the rat's forelimb in different postures they could alter the apparent map of muscles in cortex. For example, when the forelimb was in an extended posture, the biceps representation in cortex was enlarged. When the forelimb was in a flexed posture, the biceps representation in cortex shrank. This type of change in the cortical representation of muscles due to proprioceptive feedback has been obtained in many experiments in humans, monkeys, and cats (Armstrong & Drew 1985; Bennett & Lemon 1994; Graziano et al. 2004; Lemon et al. 1995; Rho et al. 1999).

Figure 1A shows an example from a recent experiment (Graziano et al. 2004) in which proprioceptive information about the angle of the elbow joint altered

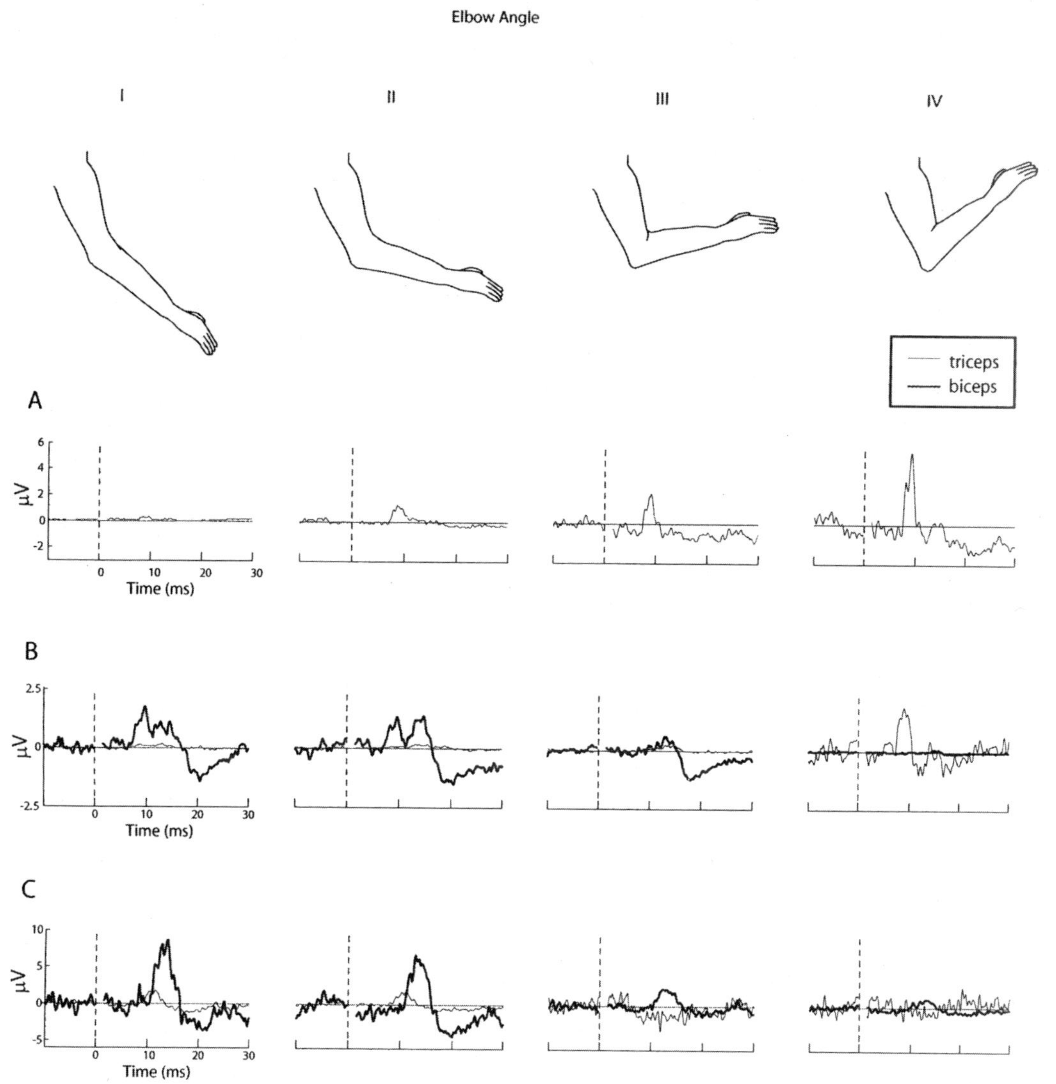

FIGURE 1. Cortico-muscle connectivity modulated by proprioceptive feedback. Top: The arm was fixed in four possible locations in an anesthetized monkey while biphasic stimulation pulses were applied to points in cortex (30 microamps, 15 Hz, 0.2 ms width per phase, negative phase leading). Electromyographic (EMG) activity was recorded in biceps and triceps. A. EMG activity in triceps evoked by stimulation of one point in primary motor cortex. Vertical line on each histogram indicates time of biphasic pulse delivered to brain. Time from 0.2 ms before to 1.5 ms after the pulse is removed from the EMG data to avoid electrical artifact. Each histogram is a mean of 2000–4500 pulses. The stimulation-evoked activity was modulated by the angle of the joint. Thus the effective connection strength between the stimulated point in cortex and the muscle was modulated by joint angle. B. EMG activity in biceps and triceps evoked by stimulation of a second example point in primary motor cortex. Stimulation of this point in cortex could activate the biceps or the triceps depending on the angle of the joint. One interpretation is that activity at that location in cortex signals the elbow to move from any initial angle toward an intermediate, final angle. When the elbow is more flexed than the desired final angle, stimulation evokes mainly triceps activity. When the elbow is more extended than the desired final angle, stimulation evokes mainly biceps activity. C. EMG activity in biceps and triceps evoked by stimulation of a third example point in primary motor cortex. Stimulation of this point in cortex activated primarily the biceps. One interpretation is that activity at that location in cortex signals the elbow to move in a controlled fashion toward flexion. When the elbow is far from a flexed position, stimulation evokes a higher level of biceps activity and a greater discrepancy between biceps and triceps activity. When the elbow is near full flexion, stimulation evokes a lower level of biceps activity and a smaller discrepancy between biceps and triceps activity.

the effective connectivity between a point in cortex and the triceps. Here we collected data from an anesthetized monkey whose elbow was fixed at several different angles. Stimulation pulses applied to this site in cortex resulted in a short latency activation of the triceps. The amount of triceps activation was modulated in a simple, monotonic, roughly linear fashion by the angle at which the elbow joint was fixed. The more flexed the elbow, the greater the evoked muscle activity.

It is important to note that the change in evoked activity in the triceps was not a result of a length/tension relationship, in which muscle tension varies with muscle length due to the physical properties of the muscle. Here we were not measuring the evoked tension in the muscle, but the evoked electromyographic activity.

It is also important to note that, by using the technique of stimulus triggered averages (Cheney et al. 1985), the experiment was able to probe a short-latency (approximately 7 ms) neuronal pathway from the stimulated site in cortex to the muscle. The modulation caused by elbow angle must have occurred along this relatively direct pathway. The proprioceptive feedback could have modulated various steps along this pathway, such as altering the stimulation threshold of the neurons in cortex near the electrode tip, altering the circuitry within the spinal cord, or both. For example, stretch receptors in the biceps and triceps might have fed back to the spinal cord and altered the excitability of the alpha motor neuron pool for the triceps.

The example in Figure 1A represents a relatively simple building block, a cortico-muscle connection that is modulated in a monotonic, roughly linear fashion by joint angle. In the following sections we discuss how this simple building block might be used to control highly complex movement parameters.

Remapping a Point in Cortex from Flexor to Extensor

Figure 1B shows an example in which a point in motor cortex was remapped from the biceps to the triceps when the elbow angle was changed (Graziano et al. 2004). Here we stimulated a point in motor cortex and found a short-latency excitatory response in both the biceps and triceps. When the elbow was fixed in an extended posture, activity at that point in cortex excited the biceps more than the triceps. When the elbow was fixed in a flexed posture, activity at that point in cortex excited the triceps more than the biceps. Essentially, this point in cortex could be functionally connected to the flexors or to the extensors depending on the angle of the elbow.

Our interpretation in the present example is that the pattern of activity is designed to initiate movement of the elbow toward an intermediate, goal angle, regardless of the starting angle. When the arm is initially extended, the increase in biceps activity should initiate a flexion. When the arm is initially flexed, the increase in triceps activity should initiate an extension. Indeed, when this site in cortex was stimulated with a 400-ms train of pulses presented at 200 Hz, and the arm was free to move, the elbow moved to a partially flexed angle regardless of its starting angle and then remained at that final posture until the end of the stimulation train.

In this interpretation, the output neurons at the stimulated site in cortex did not encode a specific pattern of muscle activity; instead, they encoded movement to a desired posture. Thus a fundamentally muscle-based map, with the addition of a simple feedback remapping rule, can in principle be used to construct a higher-order, postural code for movement.

Movement to an Extreme Angle

As described above, for some sites in cortex, stimulation can result in movement of a joint to a goal angle. For other sites, however, stimulation results in movement of a joint in one direction only. If such a site in cortex is stimulated for a long enough duration, the joint reaches an extreme position. This type of site was classically described with respect to the control of the fingers (Asanuma 1975). This pattern of results was interpreted as evidence of a relatively direct, fixed connection between the stimulated point in cortex and a single muscle, either a flexor or an extensor. However, even in this case, the mapping between cortex and muscle may not be simple or fixed and may be modulated by proprioceptive feedback.

Figure 1C shows an example of a site in cortex that when stimulated always drove the elbow toward flexion (Graziano et al. 2004). The evoked muscle activity was nonetheless modulated by joint angle. In this case, the strength of the cortico-biceps pathway was greatest when the elbow was fully extended and least when the elbow was fully flexed. The discrepancy between biceps and triceps activity was also greatest the elbow was extended and least when the elbow was flexed.

The practical effect of this modulation is that activity at this site should initiate a regulated movement of the elbow toward flexion, in which the amount of muscle activity depends on how far the elbow must be moved to reach full flexion. In this interpretation, the output neurons at the stimulated point in cortex did not encode a fixed pattern of muscle activity. Instead, they encoded a regulated *movement* toward flexion in which different patterns of muscle activity might be required under different circumstances. At other stimulation sites, for which stimulation resulted in an extension of the elbow, a corresponding result was obtained with respect to the triceps.

Feedback Remapping and the Coding of Movement Direction

In principle, the same mechanism of feedback remapping outlined above could allow proprioceptive feedback from one joint to modulate the connections between cortex and the muscles that cross a different joint. In this way, the movement or position of one joint could interact with the cortical control of another joint. One example of this type of feedback remapping was provided by Kakei et al. (1999). They recorded from neurons in the motor cortex of monkeys performing a wrist movement task. For some neurons, the orientation of the forearm remapped the relationship between neuronal activity and the muscles that actuate the wrist. For example, for one type of neuron, if the forearm was supinated (palm up), activity of the neuron was correlated with, and presumably helped to drive, the muscles that flex the wrist, resulting in the hand rotating upward. If the forearm was pronated (palm down), activity of the neuron was correlated with the muscles that extend the wrist, again resulting in the hand rotating upward. In this example, a single neuron in cortex encoded "upward" movement of the wrist regardless of the orientation of the limb. The underlying computation is the same as in the example in Figure 1B. In both cases, a point in cortex was connected primarily to the flexors or to the extensors depending on feedback about the angle of a joint. In the example from Kakei et al., the remapping resulted in a code for direction of movement in extrinsic space.

Feedback remapping could in principle be used to construct other complex codes for movement as well. For example, dynamic stretch receptors in the muscles detect the speed of joint rotation, and therefore could modulate the mapping from cortex to muscles on the basis of velocity, resulting in a movement code in which neurons in cortex help to specify the velocity of the movement (e.g. Reina et al. 2001). Feedback remapping could also result in combinations of different types of coding, in which aspects of posture, direction, and speed are all controlled to some degree to result in a complex action. Such actions that appear to combine the control of many different parameters are reminiscent of the movements evoked by electrical stimulation of motor cortex, such as bringing the hand to the mouth in an apparently speed-controlled manner (Graziano et al. 2002a,b).

Summary

A traditional debate in motor physiology is whether motor cortex controls behavior at the level of movements or of muscles (Taylor & Gross 2003). Neurons in motor cortex become active in correlation with many movement parameters such as direction of movement of the hand through space, velocity, force, joint angle, and arm posture (e.g. Caminiti et al. 1990; Evarts 1968; Georgopoulos et al. 1986; Georgopoulos et al. 1989; Kakei et al. 1999; Kalaska et al. 1989; Reina et al. 2001; Scott& Kalaska 1995; Scott & Kalaska 1997; Sergio & Kalaska 2003). Electrical stimulation of motor cortex on a behavioral time scale results in complex, multijoint movements that appear to match the monkey's normal behavioral repertoire (Cooke & Graziano 2004; Graziano et al. 2002a,b; Graziano et al. 2004). Even purely spatial information separated from any overt movement can influence neurons in motor cortex (Crowe et al. 2004). It is therefore clear that motor cortex is not simply a topographic map of muscles. Yet it does have a relatively direct, descending pathway to the muscles, and neurons in motor cortex are highly correlated with muscle output (Cheney et al. 1985; Evarts 1968; He et al. 1993; Holdefer & Miller 2002; Lemon et al. 1986). Perhaps the relevant question is not whether motor cortex controls muscles or movements, since it clearly does both. Rather, the relevant question may be: what are the variables that intervene between motor cortex and muscles?

Here we emphasize that proprioceptive feedback from the limb is an important class of variables that intervenes between motor cortex and muscles. In this view, motor cortex is mapped to muscles, and this mapping can be changed on a moment-by-moment basis as a result of feedback from joint angle and muscle stretch. We propose that this feedback remapping provides tremendous processing power and can underlie the cortical control of both simple and complex motor variables, such as when activity in cortex specifies a flexion or extension of a joint, a goal angle for a joint, a movement in a particular direction in space, or a movement of a particular peak speed. We suggest that feedback remapping may be an overarching method of motor control that can be used to construct many different, specific motor coding schemes. These specific motor coding schemes might depend on the subregion of motor cortex under study, the body part being controlled, the task being performed by the animal, or the training history of the animal.

References

Armstrong DM, Drew T (1985) Forelimb electromyographic responses to motor cortex stimulation during locomotion in the cat. J Physiol 367: 327–351.

Asanuma H (1975) Recent developments in the study of the columnar arrangement of neurons within the motor cortex. Physiol Rev 55: 143–156.

Asanuma H, Zarzecki P, Jankowska E, Hongo T, Marcus S (1979) Projection of individual pyramidal tract neurons to lumbar motor nuclei of the monkey. Exp Brain Res 34: 73–89.

Baker SN, Olivier E, Lemon RN (1998) An investigation of the intrinsic circuitry of the motor cortex of the monkey using intra-cortical microstimulation. Exp Brain Res 123: 397–411.

Bennett KM, Lemon RN (1994) The influence of single monkey cortico-motoneuronal cells at different levels of activity in target muscles. J Physiol 477: 291–307.

Bizzi E, Tresch MC, Saltiel P, d'Avella A (2000) New perspectives on spinal motor systems. Nat Rev Neurosci 1: 101–108.

Caminiti R, Johnson PB, Urbano A (1990) Making arm movements within different parts of space: dynamic aspects in the primate motor cortex. J Neurosci 10: 2039–2058.

Capaday C, Devanne H, Bertrand L, Lavoie BA (1998) Intracortical connections between motor cortical zones controlling antagonistic muscles in the cat: a combined anatomical and physiological study. Exp Brain Res 120: 223–232.

Cheney PD, Fetz EE (1985) Comparable patterns of muscle facilitation evoked by individual corticomotoneuronal (CM) cells and by single intracortical microstimuli in primates: evidence for functional groups of CM cells. J Neurophys 53: 786–804.

Cheney PD, Fetz EE, Palmer SS (1985) Patterns of facilitation and suppression of antagonist forelimb muscles from motor cortex sites in the awake monkey. J Neurophys 53: 805–820.

Cooke DF, Graziano MSA (2004) Sensorimotor integration in the precentral gyrus: polysensory neurons and defensive movements. J Neurophys 91: 1648–1660.

Crowe DA, Chafee MV, Averbeck BB, Georgopoulos AP (2004) Participation of primary motor cortical neurons in a distributed network during maze solution: representation of spatial parameters and time-course comparison with parietal area 7a. Exp Brain Res 158: 28–34.

Donoghue JP, LeiBovic S, Sanes JN (1992) Organization of the forelimb area in squirrel monkey motor cortex: representation of digit, wrist, and elbow muscles. Exp Brain Res 89: 1–19.

Evarts EV (1968) Relation of pyramidal tract activity to force exerted during voluntary movement. J Neurophys 31: 14–27.

Fetz EE, Cheney PD (1980) Postspike facilitation of forelimb muscle activity by primate corticomotoneuronal cells. J Neurophys 44: 751–772.

Foerster O (1936) The motor cortex of man in the light of Hughlings Jackson's doctrines. Brain 59: 135–159.

Fulton JF (1938) Physiology of the Nervous System. Oxford University Press, New York, p. 399–457.

Gatter KC, Sloper JJ, Powell TP (1978) The intrinsic connections of the cortex of area 4 of the monkey. Brain 101: 513–541.

Georgopoulos AP, Schwartz AB, Kettner RE (1986) Neuronal population coding of movement direction. Science 233: 1416–1419.

Georgopoulos AP, Lurito JT, Petrides M, Schwartz AB, Massey JT (1989) Mental rotation of the neuronal population vector. Science 243: 234–236.

Ghosh S, Porter R (1988) Morphology of pyramidal neurones in monkey motor cortex and the synaptic actions of their intracortical axon collaterals. J Physiol 400: 593–615.

Gould HJ 3rd, Cusick CG, Pons TP, Kaas JH (1986) The relationship of corpus callosum connections to electrical stimulation maps of motor, supplementary motor, and the frontal eye fields in owl monkeys. J Comp Neurol 247: 297–325.

Graziano MSA, Taylor CSR, Moore T (2002) Complex movements evoked by microstimulation of precentral cortex. Neuron 34: 841–851.

Graziano MSA, Taylor CSR, Moore T, Cooke DF (2002) The cortical control of movement revisited. Neuron 36: 349–362.

Graziano MSA, Patel KT, Taylor CSR (2004) Mapping from motor cortex to biceps and triceps altered by elbow angle. J Neurophys 92: 395–407.

He SQ, Dum RP, Strick PL (1993) Topographic organization of corticospinal projections from the frontal lobe: motor areas on the lateral surface of the hemisphere. J Neurosci 13: 952–980.

Holdefer RN, Miller LE (2002) Primary motor cortical neurons encode functional muscle synergies. Exp Brain Res 146: 233–243.

Huntley GW, Jones EG (1991) Relationship of intrinsic connections to forelimb movement representations in monkey motor cortex: a correlative anatomic and physiological study. J Neurophys 66: 390–413.

Jankowska E, Hammar I (2002) Spinal interneurones; how can studies in animals contribute to the understanding of spinal interneuronal systems in man? Brain Res Rev 40:19–28.

Jankowska E, Padel Y, Tanaka R (1975) The mode of activation of pyramidal tract cells by intracortical stimuli. J Physiol 249: 617–636.

Kakei S, Hoffman D, Strick P (1999) Muscle and movemet representations in the primary motor cortex. Science 285: 2136–2139.

Kalaska JF, Cohen DA, Hyde ML, Prud'homme MA (1989) Comparison of movement direction-related versus load direction-related activity in primate motor cortex, using a two-dimensional reaching task. J Neurosci 9: 2080–2102.

Kang Y, Endo K, Araki T (1988) Excitatory synaptic actions between pairs of neighboring pyramidal tract cells in the motor cortex. J Neurophys 59: 636–647.

Kuang RZ, Kalil K (1990) Branching patterns of corticospinal axon arbors in the rodent. J Comp Neurol 292: 585–598.

Kwan HC, MacKay WA, Murphy JT, Wong YC (1978) Spatial organization of precentral cortex in awake primates. II. Motor outputs. J Neurophys 41: 1120–1131.

Kwan HC, Murphy JT, Wong YC (1987) Interaction between neurons in precentral cortical zones controlling different joints. Brain Res 400: 259–269.

Landgren S, Phillips CG, Porter R (1962) Cortical fields of origin of the monosynaptic pyramidal pathways to some alpha motoneurones of the baboon's hand and forearm. J Physiol 161: 112–125.

Landry P, Labelle A, Deschenes M (1980) Intracortical distribution of axonal collaterals of pyramidal tract cells in the cat motor cortex. Brain Res 191: 327–336.

Lemon RN, Johansson RS, Westling G (1995) Corticospinal control during reach, grasp, and precision lift in man. J Neurosci 15: 6145–6156.

Lemon RN, Kirkwood PA, Maier MA, Nakajima K, Nathan P (2004) Direct and indirect pathways for corticospinal control of upper limb motoneurons in the primate. Prog Brain Res 143: 263–279.

Lemon RN, Mantel GW, Muir RB (1986) Corticospinal facilitation of hand muscles during voluntary movement in the conscious monkey. J Physiol 381: 497–527.

Maier MA, Armand J, Kirkwood PA, Yang HW, Davis JN, Lemon RN (2002) Differences in the corticospinal projection from primary motor cortex and supplementary motor area to macaque upper limb motoneurons: an anatomical and electrophysiological study. Cereb Cortex 12: 281–296.

Maier MA, Olivier E, Baker SN, Kirkwood PA, Morris T, Lemon RN (1997) Direct and indirect corticospinal control of arm and hand motoneurons in the squirrel monkey (Saimiri sciureus). J Neurophys 78: 721–733.

Matsumura M, Chen D, Sawaguchi T, Kubota K, Fetz EE (1996) Synaptic interactions between primate precentral cortex neurons revealed by spike-triggered averaging of intracellular membrane potentials in vivo. J Neurosci 16: 7757–7767.

McKiernan BJ, Marcario JK, Karrer JH, Cheney PD (1998) Corticomotoneuronal postspike effects in shoulder, elbow, wrist, digit, and intrinsic hand muscles during a reach and prehension task. J Neurophys 80: 1961–1980.

Murray EA, Coulter JD (1981) Organization of corticospinal neurons in the monkey. J Comp Neurol 195: 339–365.

Morrow MM, Miller LE (2003) Prediction of muscle activity by populations of sequentially recorded primary motor cortex neurons. J Neurophys 89: 2279–2288.

Olivier E, Baker SN, Nakajima K, Brochier T, Lemon RN (2001) nvestigation into non-monosynaptic corticospinal excitation of macaque upper limb single motor units. J Neurophys 86: 1573–1586.

Park MC, Belhaj-Saif A, Gordon M, Cheney PD (2001) Consistent features in the forelimb representation of primary motor cortex in rhesus macaques. J Neurosci 21: 2784–2792.

Reina GA, Moran DW, Schwartz AB (2001) On the relationship between joint angular velocity and motor cortical discharge during reaching. J Neurophys 85: 2576–2589.

Rho MJ, Lavoie S, Drew T (1999) Effects of red nucleus microstimulation on the locomotor pattern and timing in the intact cat: a comparison with the motor cortex. J Neurophys 81: 2297–2315.

Sanes JN, Donoghue JP, Thangaraj V, Edelman RR, Warach S (1995) Shared neural substrates controlling hand movements in human motor cortex. Science 268: 1775–1777.

Sanes JN, Wang J, Donoghue JP (1992) Immediate and delayed changes of rat cortical output representation with new forelimb configurations. Cereb Cortex 2: 141–152.

Schieber MH, Hibbard LS (1993) How somatotopic is the motor cortex hand area? Science 261: 489–492.

Schneider C, Devanne H, Lavoie BA, Capaday C (2002) Neural mechanisms involved in the functional linking of motor cortical points. Exp Brain Res 146: 86–94.

Schneider C, Zytnicki D, Capaday C (2001) Quantitative evidence for multiple widespread representations of individual muscles in the cat motor cortex. Neurosci Lett 310: 183–187.

Scott SH, Kalaska JF (1995) Changes in motor cortex activity during reaching movements with similar hand paths but different arm postures. J Neurophys 73: 2563–2567.

Scott SH, Kalaska JF (1997) Reaching movements with similar hand paths but different arm orientations. I. Activity of individual cells in motor cortex. J Neurophys 77: 826–852.

Sergio LE, Kalaska JF (2003) Systematic changes in motor cortex cell activity with arm posture during directional isometric force generation. J Neurophys 89: 212–228, 2003.

Shinoda Y, Arnold AP, Asanuma H (1976) Spinal branching of corticospinal axons in the cat. Exp Brain Res 26: 215–234.

Shinoda Y, Yamaguchi T, Futami T (1986) Multiple axon collaterals of single corticospinal axons in the cat spinal cord. J Neurophys 55: 425–448.

Tantisira B, Alstermark B, Isa T, Kummel H, Pinter M (1996) Motoneuronal projection pattern of single C3-C4 propriospinal neurones. Can J Physiol Pharmacol 74: 518–530.

Taylor CSR, Gross, CG (2003) Twitches vs. Movements: a story of motor cortex. Neuroscientist 9: 332–342.

Todorov E (2000) Direct cortical control of muscle activation in voluntary arm movements: a model. Nat Neurosci 3: 391–398.

Todorov E, Jordan MI (2002) Optimal feedback control as a theory of motor coordination. Nat Neurosci 5: 1226–1235.

Weiss DS, Keller A (1994) Specific patterns of intrinsic connections between representation zones in the rat motor cortex. Cereb Cortex 4: 205–214.

10. HOW CEREBRAL AND CEREBELLAR PLASTICITIES MAY COOPERATE DURING ARM REACHING MOVEMENT LEARNING: A NEURAL NETWORK MODEL

Alexander A. Frolov

Institute of Higher Nervous Activities and Neurophysiology, Russian Academy of Sciences, Butlerova 5a 117 485 Moscow, Russia. E-mail: aafrolov@mail.ru

Michel Dufossé

INSERM U483, Université Pierre and Marie Curie, CP-23, 75005 Paris, France. E-mail: michel.dufosse@snv.jussieu.fr

Abstract

Learning process results from synaptic plasticities that occur in various sites of the brain. For arm reaching movement, three sites have been particularly studied: the cortico-cortical synapses of the cerebral cortex, the parallel fibre-Purkinje cell synapses of the cerebellar cortex and the cerebello-thalamo-cortical pathway. We intended to understand how these three adaptive processes cooperate for optimal performance. A neural network model was developed based on two main prerequisites: the columnar organisation of the cerebral cortex and the Marr-Albus-Ito theory of cerebellar learning. The adaptive rules incorporated in the model simulate the synaptic plasticities observed at the three sites. The model analytically demonstrates that 1) the adaptive processes that take place in different sites of the cerebral cortex and the cerebellum do not interfere but complement each other during learning of arm reaching movement, and 2) any linear combination of the cerebral motor commands may generate olivary signals able to drive the cerebellar learning processes.

Keywords: motor learning, plasticity, cerebellum, inferior olive, cerebro-cerebellar interaction.

1. *Introduction*

The control process of reaching a visual target with the hand is performed by a transformation from visual and parietal to motor cortex through two main pathways, a direct cerebral one and a cerebellar side-path. During learning, plasticity mainly occurs at three sites, the cortico-cortical synapses, the parallel fibre-Purkinje cell synapses and the cerebello-thalamo-cortical pathway at the thalamo-cortical synapses. Our goal was to understand how the three adaptive processes cooperate for an optimal performance.

A neural network model was developed based on two prerequisites. The first one is the theory of a cerebral operating mode through local populations of neurones located perpendicularly to the cortical surface (Arbib et al., 1988; Burnod, 1989). At different layers of these columnar-like populations, different information processes interact and several synaptic plasticities are observed.

The second prerequisite is the Marr-Albus-Ito theory of cerebellar learning (Marr, 1969; Albus, 1971; Ito, 1984). Learning is supervised by the inferior olive nucleus which sends one climbing fibre to each Purkinje cell, the principal cell of the cerebellar cortex. The effect of climbing fibre activity on a Purkinje cell is a long term depression of the synaptic strength of the parallel fibres whose activities are correlated with it. As a result, the learning process was interpreted as an error reduction. However, the generation of climbing fibre error signals raises two main questions. In the temporal domain, any error in performance occurs far too late for being used as a supervising error signal. In the spatial (somatotopic) domain, it is unlikely that the inferior olivary nucleus may compute all the specific error signals that the fifteen millions of Purkinje cells need in the human, depending of the motor synergies to which they contribute.

From the target position in the extracorporeal space, the neural network computes the motor commands acting on the executive organ (the plant). Thus, it performs the function of an inverse model of the plant. Several general schemes have been previously proposed for learning such inverse model. They mainly differ by the way of how is delivered the error signal able to supervise learning. The idea of inverse modelling (for example, Kuperstein, 1988) arises from observing the input/output behavior of the executive organ, and training an inverse model by reversing the roles of input and output. However, this scheme is not applicable to a redundant executive organ, where the same movement can be produced by a multitude of motor commands. The idea of forward-inverse modelling (Jordan, Rumelhart, 1992) is to train an inverse model by means of a previously learned forward model. However, this scheme requires an physiologically unfeasible mechanism of error back propagation through the forward model. In addition, both schemes require an physiologically unfeasible switching of input signals between the operative and learning modes. Thus, a feedback error learning (Kawato et al., 1987) was proposed, that can train the inverse model by the error signals produced by a feedback controller. This scheme is physiologically more plausible, but it requires a pre-existing accurate feedback controller. This drawback of feedback error learning has been overcome in (Frolov, Rizek, 1995), where the feedback controller is taught to be proportional to the transposed Jacobian of the executive organ. In this case, the learning of feedback controller is based on unsupervised Hebbian rule, and can be performed before, or in parallel with, the learning of an inverse model. Another scheme of feedback controller learning has been proposed in (Oyama et al., 2001), where the feedback controller is taught to be proportional to the pseudo-inverse Jacobian of the executive organ.

The cortical part of the model presented in this study exploits the same scheme as in (Frolov, Rizek, 1995), while cerebellar learning is based on a quite new idea. The signal required to supervise cerebellar learning is not related to an error in motor performance. Moreover, this signal is generated before the movement starts and thus, before any error in performance can be detected. The cerebellar learning results from successful interaction between the three neural plasticities mentioned above.

The mathematical analysis of the interaction between the learning rules requires the model to be linear. However this restriction is not essential, since the nonlinear properties of the executive organ can be taken into account by various approaches, such as the use of multi-layer nonlinear network with learning by error back propagation (Jordan, Rumelhart, 1992), the use of additional input to the linear neural network by nonlinear transformations of plant coordinates (Kawato et al., 1987), the combining of the visual information on target position with the proprioceptive information on arm position to adjust the linear visuo-motor transformation to the current arm position (Baraduc et al., 2001) or the division of the operational space into many subspaces, within which the plant can be approximated as linear (Frolov, Rizek, 1995).

As another simplification, the dynamic properties of the plant are ignored. This simplification follows from equilibrium point theory (Feldman, 1966) which assumes that the brain controls a simple planned trajectory of an equilibrium position or skeletal configuration. The control system must only solve the static inverse problem in order to control a movement. Due to the skeleto-muscular inertial and viscoelastic parameters, muscle forces result from the difference between the actual and virtual trajectories, that tend to separate, for example when the limb acceleration is large or when external forces are applied. Simulation of a two links model equipped with six muscles confirmed the ability of static inverse model to control the movement of such dynamical system as human arm (Gribble et al., 1998; Frolov et al., 2000).

The mathematical analysis of the model demonstrates that the adaptive processes that take place in different sites of the cerebral cortex and the cerebellum do not interfere but cooperate. The cerebral learning tends to lead the cerebellar learning. In addition, it is shown that any linear combination of the cerebral motor commands may generate olivary signals able to supervise the cerebellar learning processes. These early signals arrive in conjunction with the context of the on-going action, reflected by the parallel fiber activities.

2. *Control Model*

As generally accepted, we assume that the brain controls a working point (WP), such as the finger tip in a reaching movement, planning its movement from an initial to a final position with the aim of reaching a target. The desired WP displacement is internally represented in the parietal cortex by a set of cell activities which code the projections of this displacement onto their preferred directions in space (Georgopoulos et al., 1984).

In turn, the actual WP displacement is internally represented in the motor cortex by a set of cell activities which code the contributions of these cells to the motor execution. The direction of action of a cell is defined as the vector of elementary motor action which would be produced by its activation. Then, the activity of each cell contributes to the WP displacement along its direction of action. To produce a desired WP displacement, the central nervous system must perform the visuomotor transformation of the visual target representation to the motor commands.

Two main pathways allow the brain to perform this transformation. In a cerebral pathway (Fig. 1, top), a first matrix **M**, with fixed random coefficients, performs the projection of the vector target (**T**) on the preferred directions (the row-vectors of matrix **M**) of a set of parietal columns. These directions are uniformly distributed in space (Georgopoulos et al., 1984). This matrix provides an internal visual representation of the target **T** in the form $\mathbf{A}_0 = \mathbf{MT}$. Each coefficient of vector $\mathbf{A}_0$ represents the activity of a column of the parietal cortex.

Another matrix **U** with adaptive coefficients performs the crude visuomotor transformation of the target, providing an internal motor target representation in the form $\mathbf{A}_1 = \mathbf{UA}_0$. The third matrix $\mathbf{V}_1$ whose coefficients are also adaptive, performs the correction of the internal motor target presentation ($\mathbf{A}_1$) in the vector of neuronal activities $\mathbf{A}_2 = \mathbf{V}_1\mathbf{A}_1$. Both signals $\mathbf{A}_1$ and $\mathbf{A}_2$ contribute to the motor command **C**.

In the cerebellar side-pathway (Fig. 1, bottom), the matrix of projection **N** provides another internal visual target representation in terms of the parallel fibre activities, the vector $\mathbf{B}_0 = \mathbf{NT}$. The first adaptive matrix $\mathbf{V}_2$ performs the parallel fibre/Pukinje cell transformation, which provides the vector of Purkinje cell activities $\mathbf{B}_1 = \mathbf{V}_2\mathbf{B}_0$. The second adaptive matrix $\mathbf{V}_3$ performs the cerebello-thalamo-cortical transformation, which in turn provides another internal motor representation of the target in terms of the vector of cerebral activities $\mathbf{B}_2 = \mathbf{V}_3\mathbf{B}_1$.

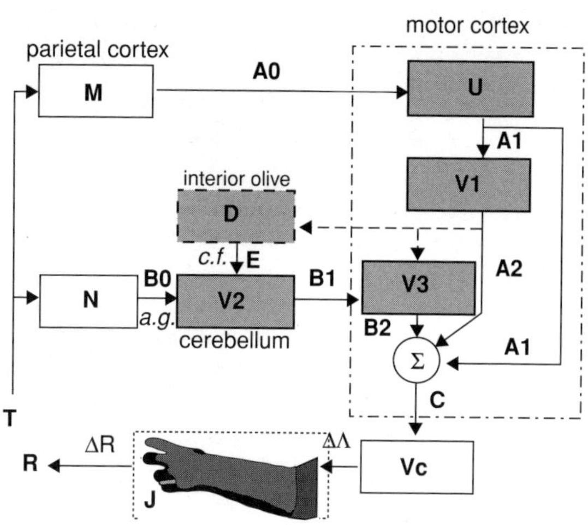

FIGURE 1. The model of the sensorimotor transformation. Two pathways, cerebral (top) and cerebellar (bottom), link a visual target (**T**) to the motor command (**C**) leading to arm position (**R**). Neuronal activities are represented by the vectors: $\mathbf{A}_0$ for the parietal cortex, $\mathbf{A}_1$, $\mathbf{A}_2$ and $\mathbf{B}_2$ for the motor cortex, $\mathbf{B}_0$ for cerebellar granular cells, $\mathbf{B}_1$ for Purkinje cells, and **E** for climbing fibres. Boxes represent matrices of synaptic weights. Four of them have adaptive values: **U** and $\mathbf{V}_1$ for cerebral cortico-cortical synapses, $\mathbf{V}_2$ for parallel fibres/Pukinje cell synapses and $\mathbf{V}_3$ for cerebello-thalamo-cortical synapses.

Vectors $\mathbf{A}_1$, $\mathbf{A}_2$ and $\mathbf{B}_2$ are summed into the command signal $\mathbf{C} = \mathbf{A}_1 + \mathbf{A}_2 + \mathbf{B}_2$, which determines the change of the hand equilibrium position in the external space. Components of vectors $\mathbf{A}_1$, $\mathbf{A}_2$, $\mathbf{B}_2$ and **C** with the same indices represent neuronal activities at different layers of the same cerebral column.

The transformation of the command signal **C** into the final arm position **R** is performed through the pathway from motor cortex to the neuromuscular apparatus of the executive system (the pyramidal tract). According to the equilibrium point theory, this command vector **C** is transformed into the control vector $\Delta\mathbf{\Lambda}_c$ which is the change of the supraspinal input to alpha-motoneurones controlling muscle forces and sequentially hand position. Vector $\Delta\mathbf{\Lambda}_c$ is summed with the initial vector $\mathbf{\Lambda}_c^{in}$ creating the actual input to alpha-motoneurones $\mathbf{\Lambda}_c = \mathbf{\Lambda}_c^{in} + \Delta\mathbf{\Lambda}_c$. Each component of $\mathbf{\Lambda}_c$ determines the static force of one individual muscle. The equilibrium state is achieved when the forces produced by antagonistic muscles are mutually compensated. The change of $\mathbf{\Lambda}_c$ produces the change of muscle forces and consequently the change of equilibrium arm position, i.e. the arm movement.

We assume that transformation of **C** into $\Delta\mathbf{\Lambda}_c$ is linear and performed by matrix $\mathbf{V_c}$ in the form $\Delta\mathbf{\Lambda}_c = \mathbf{V}_c\mathbf{C}$. The nonlinearity of the visuomotor transformation is only related to the nonlinearity of the transformation of $\mathbf{\Lambda}_c$ into muscle forces and nonlinearity of transformation of muscle forces into WP position **R**. We do not treat here the problem of nonlinearity assuming that it can be overcome by methods discussed above. The change of WP position ($\Delta\mathbf{R} = \mathbf{R} - \mathbf{R}_{in}$, **R** and $\mathbf{R}_{in}$ being the actual and initial WP positions) is given by $\Delta\mathbf{R} = \mathbf{J}\Delta\mathbf{\Lambda}_c$ where **J** is the Jacobian of the plant that is assumed to be constant in linear approximation. As a whole, the command signal **C** is transformed into WP displacement $\Delta\mathbf{R}$ by the output matrix $\mathbf{O} = \mathbf{J}\mathbf{V}_c$. Dimensionality of the control signal space (vector $\mathbf{\Lambda}_c$) and command signal space (vector **C**) are larger than the dimensionality of the working space. Thus the problem of visuo-motor transformation is ill-posed and cooperating neural networks solve it specifically according to their architecture and learning rules.

3. *Learning Procedures*

3.1. SITES OF PLASTICITY

Several sites of plasticity are known in the cerebro-cerebellar network: the cerebral cortico-cortical connections, the cerebellar synapses from parallel fibres to Purkinje cells and the cerebello-thalamo-cortical pathway. The first and third learning rules are here considered to be unsupervised Hebbian rules. By contrast, the cerebellar learning rule is supervised by the inferior olive nucleus, which activity is governed by cortical signals (here the contribution of peripheral structures to their activity is not taken into account).

Cerebellar learning was hypothesized to be located at the parallel fibre contacts with the Purkinje cells (Marr, 1969; Albus, 1971). Experiments in alert animals (Gilbert and Thach, 1977; Dufossé et al., 1978) and in anaesthetized preparation (Ito et al., 1982) confirmed the hypothesis. The repeated conjunction of climbing and parallel fibre activities on a Purkinje cell produces a long-term depression in the synaptic strength of the parallel fibre synapse (Daniel et al., 1998). In turn, the repetitive firing of parallel fibres alone can reverse the process by inducing long-term potentiation (Crépel and Jaillard, 1991). These opposing tendencies probably contribute to a distribution of the strengths of parallel fibres synapses appropriate to an unsaturated level of Purkinje cell activity, that can be modulated. The cerebellar plasticity can be interpreted as an error learning rule, in the sense that the more a climbing fibre discharges, the less the Purkinje cell will discharge, for the same level of its parallel fibre activities. This definition, made at the cellular level, refers to a 'local error'. By contrast, the term of error was often referred to a 'functional error', such as a mismatch between the final WP position and the target during reaching or between intended and actual movement (Ito, 1984; Schweighofer et al., 1998).

Arm movements involve the intermediate and lateral parts of the cerebellum which mainly project to the cerebral cortex via the thalamus. The combined activation of thalamic input and cerebral cortico-cortical connections on a cerebral site modifies the synaptic strengths of both pathways (Baranyi and Feher, 1978; Iriki et al., 1989). The cerebello-thalamo-cortical pathway has been experimentally studied and its plasticity confirmed (Meftah and Rispal-Padel, 1992; Pananceau et al., 1996).

3.2. 'BABBLING LEARNING' (U)

The so-called 'babbling learning' process occurs when the final position of random movements is considered as being the target, and is associated with the corresponding motor commands. This learning process is assumed to be performed before all the other learning processes ($\mathbf{V}_1 = \mathbf{V}_2 = \mathbf{V}_3 = 0$).

During learning, random vectors of columnar activities $\mathbf{A}_1$ generated in the motor cortex produce motor commands $\mathbf{C} = \mathbf{A}_1$ and subsequent hand displacements $\Delta\mathbf{R} = \mathbf{J}\mathbf{V}_c\mathbf{A}_1$ which are in turn perceived in the parietal cortex as vectors $\mathbf{A}_0 = \mathbf{M}\Delta\mathbf{R}$. Both vectors are associated in the learning process by the Hebbian rule

$$\Delta\mathbf{U} = \epsilon_0\mathbf{A}_1\mathbf{A}_0^T = \epsilon_0\mathbf{A}_1(\mathbf{M}\mathbf{J}\mathbf{V}_c\mathbf{A}_1)^T,$$

where ϵ_0 is the learning rate and upper index T refers to matrix transposition. Since vectors $\mathbf{A}_1$ are assumed to be uniformly distributed in the internal motor space, $\langle\mathbf{A}_1\mathbf{A}_1^T\rangle$ tends to the matrix $k_1\mathbf{I}_c$, where $\langle\cdot\rangle$ denotes time averaging and $\mathbf{I}_c$ denotes the unit matrix in the internal motor space. Then,

$$\mathbf{U} \simeq \eta_0\mathbf{V}_c^T\mathbf{J}^T\mathbf{M}^T, \qquad (1)$$

with $\eta_0 = \epsilon_0 k_1$.

Since the preferred directions are assumed to be uniformly distributed in the external space (Georgopoulos et al., 1984), and the number of columns is much larger than the dimensionality of the external space, then $\mathbf{M}^T\mathbf{M} \simeq k_2\mathbf{I}_e$ where k_2 is a coefficient whose value depends on the number of parietal columns and $\mathbf{I}_c$ is the unit matrix in the external working space. Thus, the matrix $\mathbf{M}^* = \mathbf{UM}$ is proportional to the transpose of matrix $\mathbf{O} = \mathbf{JV}_c$. This means that, due to motor babbling, the preferred directions of motor columns (row-vectors of matrix $\mathbf{M}^*$) coincide with their direction of actions (column-vectors of matrix $\mathbf{O}$). However, their coincidence is not sufficient to produce a precise movement to the target. The transformation represented by $\mathbf{M}^*$ is not exactly the inverse of transformation $\mathbf{O}$, and the motor representation $\mathbf{A}_1$ of the target must be corrected by means of another cerebral learning.

3.3. CEREBRO-CEREBELLAR LEARNING

Once the babbling learning is achieved, three other learning processes, respectively located at the cerebro-cerebral cortical sites ($\mathbf{V}_1$), the cerebellar parallel fibres/Purkinje cells synapses ($\mathbf{V}_2$) and the thalamo-cortical synapses ($\mathbf{V}_3$), are performed concurrently.

a) Cerebral Mismatch Learning (V_1). This block also learns by a Hebbian rule linking the activity in the motor cortex $\mathbf{A}_1 = \mathbf{UMT}$ produced by the target presentation, with the activity produced in the same cortex by the error (mismatch) in movement performance $\mathbf{A}_1^* = \mathbf{UM}(\mathbf{T} - \Delta\mathbf{R})$ where $\Delta\mathbf{R} = \mathbf{OC}$ is WP displacement provoked by motor command $\mathbf{C} = \mathbf{A}_1 + \mathbf{A}_2 + \mathbf{B}_2$. Then,

$$\Delta\mathbf{V}_1 = \epsilon_1\mathbf{A}_1^*\mathbf{A}_1^T.$$

Substituting $\mathbf{U}$ from (1), we obtain

$$\Delta\mathbf{V}_1 = \epsilon_1\eta_0^2\mathbf{V}_c^T\mathbf{J}^T\mathbf{M}^T\mathbf{M}(\mathbf{I}_e - \mathbf{X})\mathbf{TT}^T\mathbf{M}^T\mathbf{MJV}_c, \tag{2}$$

where $\mathbf{X}$ is the total transformation of the target $\mathbf{T}$ to the WP displacement $\Delta\mathbf{R}$

$$\mathbf{X} = \mathbf{JV}_c[(\mathbf{V}_1\mathbf{U} + \mathbf{U})\mathbf{M} + \mathbf{V}_3\mathbf{V}_2\mathbf{N}]. \tag{3}$$

Correct visuomotor transformation requires the final WP position to coincide with the target, and so the goal of learning in the whole system is to provide the equality $\mathbf{X} = \mathbf{I}_e$. Target directions must be uniformly distributed in the external space, so that $\langle\mathbf{TT}^T\rangle \simeq k_3\mathbf{I}_e$. Averaging is performed over a period of time large enough to provide this semiequality but small enough to prevent large change of matrix $\mathbf{V}_1$. This last condition can be provided by a proper choice of the learning rate ϵ_1. Substituting these two semiequalities into (2), the learning rule can be expressed by the differential equation

$$\frac{d\mathbf{V}_1}{dt} = \mathbf{V}_c^T\mathbf{J}^T(\mathbf{I}_e - \mathbf{X})\,\mathbf{JV}_c, \tag{4}$$

where t is a number of learning trials multiplied by scaling parameter $n = \epsilon_0^2\epsilon_1 k_1^2 k_2^2 k_3$.

By means of the singular value decomposition (Golub & Van Loan, 1996), the output matrix $\mathbf{O} = \mathbf{JV}_c$ can be presented as

$$\mathbf{O} = \mathbf{G\Lambda H}^{\mathrm{T}} \tag{5}$$

where $\mathbf{\Lambda}$ is a rectangular diagonal matrix, and $\mathbf{G}$ and $\mathbf{H}$ are the square orthonormal matrices, whose columns are respectively the eigenvectors of $\mathbf{OO}^T$ of the dimensionality of the external space and $\mathbf{O}^T\mathbf{O}$ of the dimensionality of the number of columns in the motor area. These eigenvectors create new coordinate systems in external and internal spaces. In these coordinate systems, the output transformation $\mathbf{O}$ becomes the simple diagonal transformation $\mathbf{\Lambda}$, and the total transformation of the target position to the final WP position becomes

$$\mathbf{X}' = \mathbf{G}^T\mathbf{XG}. \tag{6}$$

Presenting $\mathbf{V}_1$ as

$$\mathbf{V}_1 = \mathbf{H\Lambda}^T\mathbf{W}_1\mathbf{\Lambda H}^T, \tag{7}$$

learning of $\mathbf{V}_1$ can be described by the differential equation

$$\frac{d\mathbf{W}_1}{dt} = \mathbf{I}_e - \mathbf{X}'. \tag{8}$$

b) Cerebellar Learning (V_2). In this block, the leaning process is supervised by the vector $\mathbf{E}$ of climbing fibre activities. This signal is here considered to result from the motor cortex activities $\mathbf{A}_2$, through a fixed transformation $\mathbf{D}$ that represents the cerebro-olivary pathway. Then,

$$\Delta\mathbf{V}_2 = \epsilon_2\mathbf{EB}_0^T = \epsilon_2\mathbf{DA}_2\mathbf{B}_0^T.$$

Since $\mathbf{A}_2 = \mathbf{V}_1\mathbf{UMT}$, $\mathbf{B}_0 = \mathbf{NT}$ and $\mathbf{U}$ is given by (1),

$$\Delta\mathbf{V}_2 = \epsilon_0\epsilon_2 k_1\mathbf{DV}_1\mathbf{V}_c^T\mathbf{J}^T\mathbf{M}^T\mathbf{MTT}^T\mathbf{N}^T.$$

Substituting the semiequalities previously described for $\mathbf{M}^T\mathbf{M}$ and $\langle\mathbf{TT}^T\rangle$ the learning rule can be expressed by the differential equation

$$\frac{d\,\mathbf{V}_2}{dt} = \eta_1 \mathbf{D}\mathbf{V}_1\mathbf{V}_c^T\mathbf{J}^T\mathbf{N}^T, \tag{9}$$

with $\eta_1 = \epsilon_0\epsilon_2 k_1 k_2 k_3$.

Presenting $\mathbf{V}_2$ in the form

$$\mathbf{V}_2 = \mathbf{D}\mathbf{H}\boldsymbol{\Lambda}^T\mathbf{W}_2\mathbf{G}^T\mathbf{N}^T, \tag{10}$$

learning of $\mathbf{V}_2$ can be described by the differential equation

$$\frac{d\,\mathbf{W}_2}{dt} = \eta_1 \mathbf{W}_1\boldsymbol{\Lambda}\boldsymbol{\Lambda}^T. \tag{11}$$

c) Cerebello-Thalamo-Cortical Plasticity (V_3). Plasticity of the cerebello-thalamo-cortical pathway is located in the thalamic projections on the cerebral cortex. The learning rule is a Hebbian rule between the cerebral activity $\mathbf{A}_2$ and the signal $\mathbf{B}_1$ sent from the cerebellum via the thalamus to the intermediate layer of the cortex. Then,

$$\Delta\mathbf{V}_3 = \epsilon_3\mathbf{A}_2\mathbf{B}_1^T.$$

Since $\mathbf{A}_2 = \mathbf{V}_1\mathbf{U}\mathbf{M}\mathbf{T}$ and $\mathbf{B}_1 = \mathbf{V}_2\mathbf{N}\mathbf{T}$,

$$\begin{aligned}\Delta\mathbf{V}_3 &= \epsilon_3\mathbf{V}_1\mathbf{U}\mathbf{M}\mathbf{T}(\mathbf{V}_2\mathbf{N}\mathbf{T})^T \\ &= \epsilon_3 k_1\mathbf{H}\boldsymbol{\Lambda}\boldsymbol{\Lambda}^T\mathbf{G}^T\mathbf{M}^T\mathbf{M}\mathbf{T}\mathbf{T}^T\mathbf{N}^T\mathbf{N}\mathbf{G}\mathbf{W}_2^T\boldsymbol{\Lambda}\mathbf{H}^T\mathbf{D}^T\end{aligned} \tag{12}$$

Since the preferred directions of parallel fiber activities ($\mathbf{B}_0$) are assumed to be uniformly distributed in the external space and the number of parallel fibers is much larger than the dimensionality of the external space N_e, $\mathbf{N}^T\mathbf{N} \simeq k_4\,\mathbf{I}_e$. Substituting this semiequality and the semiequalities for $\mathbf{M}^T\mathbf{M}$ and $\langle\mathbf{T}\mathbf{T}^T\rangle$ into (12), we obtain

$$\frac{d\,\mathbf{V}_3}{dt} = \eta_2\mathbf{H}\boldsymbol{\Lambda}^T\mathbf{W}_1\boldsymbol{\Lambda}\boldsymbol{\Lambda}^T\mathbf{W}_2^T\boldsymbol{\Lambda}\mathbf{H}^T\mathbf{D}^T,$$

where η_2 takes into account coefficients ϵ_1, ϵ_2, ϵ_3, k_1, k_2, k_3, and k_4.

Presenting $\mathbf{V}_3$ as

$$\mathbf{V}_3 = \mathbf{H}\boldsymbol{\Lambda}^T\mathbf{W}_3\boldsymbol{\Lambda}\mathbf{H}^T\mathbf{D}^T \tag{13}$$

learning of $\mathbf{V}_3$ can be described by the differential equation

$$\frac{d\,\mathbf{W}_3}{dt} = \eta_2\mathbf{W}_1\boldsymbol{\Lambda}\boldsymbol{\Lambda}^T\mathbf{W}_2^T. \tag{14}$$

3.4. INTEGRATED LEARNING SYSTEM

The global learning system can be expressed through matrices $\mathbf{W}_1$, $\mathbf{W}_2$ and $\mathbf{W}_3$. According to (6), matrix $\mathbf{X}'$ relates to $\mathbf{X}$ as $\mathbf{X}' = \mathbf{G}^T\mathbf{X}\mathbf{G}$. Substituting (3), (1), (7), (10) and (13) into this equation, we obtain

$$\begin{aligned}\mathbf{X}' = \boldsymbol{\Lambda}[&\boldsymbol{\Lambda}^T\mathbf{W}_3\boldsymbol{\Lambda}\mathbf{H}^T\mathbf{D}^T\mathbf{D}\mathbf{H}\boldsymbol{\Lambda}^T\mathbf{W}_2\mathbf{G}^T\mathbf{N}^T\mathbf{N} \\ &+ k_1(\boldsymbol{\Lambda}^T\mathbf{W}_1\boldsymbol{\Lambda}\boldsymbol{\Lambda}^T\mathbf{G}^T + \boldsymbol{\Lambda}^T\mathbf{G}^T)\mathbf{M}^T\mathbf{M}]\mathbf{G}\end{aligned}$$

The number of Purkinje cells (dimension of vector $\mathbf{B}_1$) is assumed to be much larger than the number of columns in motor cortex. Then $\mathbf{D}^T\mathbf{D} \simeq k_6\mathbf{I}_c$ and similarly to the previous equations

$$\mathbf{X}' = \eta_3\boldsymbol{\Lambda}\boldsymbol{\Lambda}^T + \eta_4\boldsymbol{\Lambda}\boldsymbol{\Lambda}^T\mathbf{W}_1\boldsymbol{\Lambda}\boldsymbol{\Lambda}^T + \eta_5\boldsymbol{\Lambda}\boldsymbol{\Lambda}^T\mathbf{W}_3\boldsymbol{\Lambda}\boldsymbol{\Lambda}^T\mathbf{W}_2.$$

Thus, the system of learning equations (8, 11 and 14) takes the form of the nonlinear system of differential equations

$$\begin{aligned}\frac{d\,\mathbf{W}_1}{dt} &= \mathbf{I}_e - \eta_3\boldsymbol{\Lambda}\boldsymbol{\Lambda}^T - \eta_4\boldsymbol{\Lambda}\boldsymbol{\Lambda}^T\mathbf{W}_1\boldsymbol{\Lambda}\boldsymbol{\Lambda}^T - \eta_5\boldsymbol{\Lambda}\boldsymbol{\Lambda}^T\mathbf{W}_3\boldsymbol{\Lambda}\boldsymbol{\Lambda}^T\mathbf{W}_2 \\ \frac{d\,\mathbf{W}_2}{dt} &= \eta_1\boldsymbol{\Lambda}\boldsymbol{\Lambda}^T\mathbf{W}_1 \\ \frac{d\,\mathbf{W}_3}{dt} &= \eta_2\mathbf{W}_1\boldsymbol{\Lambda}\boldsymbol{\Lambda}^T\mathbf{W}_2^T\end{aligned} \tag{15}$$

with the initial conditions $\mathbf{W}_1(0) = \mathbf{W}_2(0) = \mathbf{W}_3(0) = 0$.

It is worth to note that matrices $\mathbf{V}_1$, $\mathbf{V}_2$ and $\mathbf{V}_3$ are learned by means of available vectors of neural activity. From equations (8), (11) and (14), the learning system results from the representation of these vectors via the target and WP positions and via the fixed matrices $\mathbf{M}$, $\mathbf{N}$, $\mathbf{O}$ and $\mathbf{E}$.

Since $\boldsymbol{\Lambda}\boldsymbol{\Lambda}^T$ is a diagonal matrix of dimensionality of the external space N_e, the system splits into N_e independent scalar systems of equations for w_{1i}, w_{2i} and w_{3i} ($i = 1 \ldots N_e$), where w_{1i}, w_{2i} and w_{3i} are the components of the diagonal matrices $\mathbf{W}_1$, $\mathbf{W}_2$ and $\mathbf{W}_3$, respectively. Each system of scalar equations has a form

$$\begin{aligned}\frac{dw_{1i}}{dt} &= 1 - \eta_3\lambda_i^2 - \eta_4\lambda_i^2 w_{1i} - \eta_5\lambda_i^4 w_{2i}w_{3i} \\ \frac{dw_{2i}}{dt} &= \eta_1\lambda_i^2 w_{1i} \\ \frac{dw_{3i}}{dt} &= \eta_2\lambda_i^2 w_{1i}w_{2i}\end{aligned} \tag{16}$$

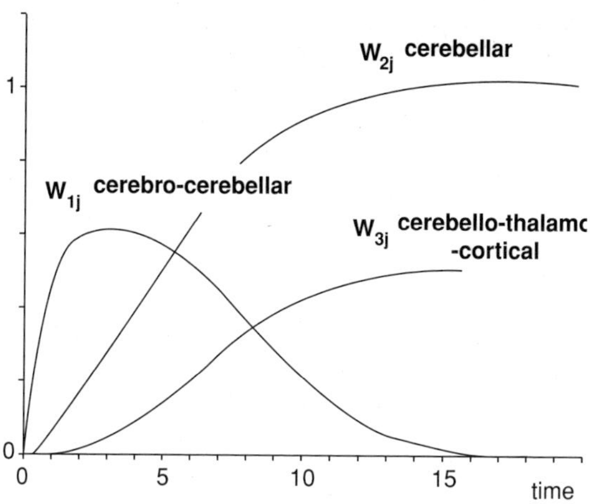

FIGURE 2. Time courses of the coefficients w_{ij} for the case $\eta_1\lambda_j^2 = 0.5$, $\eta_3\lambda_j^2 = 0.5$, $\eta_4\lambda_j^2 = 1$, and $\eta_6\lambda_j^4 = 1$. *Abscissa*: number of learning trials multiplied by scaling parameter η.

with the initial conditions $w_{1i}(0) = w_{2i}(0) = w_{3i}(0) = 0$.

From the two latter equations, $w_{3i} = \eta_2 w_{2i}^2/(2\eta_1)$; and the system takes the form

$$\frac{dw_{1i}}{dt} = 1 - \eta_3\lambda_i^2 - \eta_4\lambda_i^2 w_{1i} - \eta_6\lambda_i^4 w_{2i}^3$$

$$\frac{dw_{2i}}{dt} = \eta_1\lambda_i^2 w_{1i} \tag{17}$$

where $\eta_6 = \eta_5\eta_2/(2\eta_1)$. The solution of this system converges to $w_1 = 0$, $w_2 = [(1 - \eta_3\lambda_i^2)/(\eta_6\lambda_i^4)]^{\frac{1}{3}}$, when $t \to \infty$. When $\eta_4 > (12\eta_1\eta_6)^{\frac{1}{2}}|(1 - \eta_3\lambda^2)/(\eta_6\lambda_i)|^{\frac{1}{3}}$, the convergence is monotonic; when this condition is not satisfied, the convergence is oscillatory.

The time course of w_{ij} is shown in Fig. 2 for $\eta_1\lambda_i^2 = \eta_3\lambda_i^2 = 0.5$, $\eta_4\lambda_i^2 = \eta_6\lambda_i^2 = 1.0$, when the convergence is monotonic. It is shown that w_{1i} initially increases and then decreases. This means that during the learning process, the error signal **E** initially increases and then decreases. The parameter w_{2i} becomes larger than w_{1i}, and tends to the value $[(1 - \eta_3\lambda_i^2)/\eta_6\lambda_i^4]^{\frac{1}{3}}$, while $w_{1i} \to 0$.

Thus after cerebellar learning, the cerebral pathway no longer contributes to the control signal. However, the cerebral motor cortex still produces the crude visuo-motor transformation **U**, and receives the cerebellar signals via the thalamus. This scheme of cerebro-cerebellar interaction during motor learning was suggested by Ito (1984).

The learning system provides a solution to the problem of motor redundancy that is ill-posed, because the dimensionality of the internal motor space is much larger than the dimensionality of the external working space. It can be shown that, among the infinite number of solutions, the learning system selects the pseudo-inversion: the total transformation $\mathbf{O}^*$ of the target position $\mathbf{T}$ into the motor command $\mathbf{C}$ tends to the pseudo-inverse $\mathbf{O}^+ = \mathbf{O}^T(\mathbf{O}\mathbf{O}^T)^{-1}$ of the output transformation **O**.

According to (5)

$$\mathbf{O}^+ = \mathbf{H}\boldsymbol{\Lambda}^{\mathrm{T}}(\boldsymbol{\Lambda}\boldsymbol{\Lambda}^{\mathrm{T}})^{-1}\mathbf{G}^{\mathrm{T}}. \tag{18}$$

Since motor command is given by equation

$$\mathbf{C} = \mathbf{A}_1 + \mathbf{A}_2 + \mathbf{B}_2 = [(\mathbf{U} + \mathbf{V}_1\mathbf{U})\mathbf{M} + \mathbf{V}_3\mathbf{V}_2\mathbf{N}]\mathbf{T}$$

then

$$\mathbf{O}^* = (\mathbf{U} + \mathbf{V}_1\mathbf{U})\mathbf{M} + \mathbf{V}_3\mathbf{V}_2\mathbf{N}.$$

The substitution of **U**, $\mathbf{V}_1$, $\mathbf{V}_2$ and $\mathbf{V}_3$ into this equation, according to (1), (7), (10) and (13), gives $\mathbf{O}^* = \mathbf{H}\boldsymbol{\Lambda}^{\mathrm{T}}\mathbf{S}\mathbf{G}^{\mathrm{T}}$, with $\mathbf{S} = \eta_3\mathbf{I_e} + \eta_4\mathbf{W}_1\boldsymbol{\Lambda}\boldsymbol{\Lambda}^{\mathrm{T}} + \eta_5\mathbf{W_3}\boldsymbol{\Lambda}\boldsymbol{\Lambda}^{\mathrm{T}}\mathbf{W_2}$.

During learning, according to (15), $\boldsymbol{\Lambda}\boldsymbol{\Lambda}^T\mathbf{S} \to \mathbf{I_e}$, and then $\mathbf{S} \to (\boldsymbol{\Lambda}\boldsymbol{\Lambda}^T)^{-1}$. Consequently $\mathbf{O}^* \to \mathbf{H}\boldsymbol{\Lambda}^{\mathrm{T}}(\boldsymbol{\Lambda}\boldsymbol{\Lambda}^{\mathrm{T}})^{-1}\mathbf{G}^{\mathrm{T}} = \mathbf{O}^+$. The total transformation $\mathbf{O}^*$ of the target position **T** into the motor command **C** tends to the pseudo-inverse of the output transformation **O**.

4. *Discussion*

The model includes competitive rules of plasticity that are known to exist at different sites of the brain and shows how two brain structures, the cerebral cortex (cortico-cortical connections $\mathbf{V}_1$) and the cerebellar cortex (parallel fibres/Purkinje cells connections $\mathbf{V}_2$), may learn in parallel. In order to explain how these two structures interact during learning, the model stresses the plasticity of the cerebello-thalamo-cortical pathway ($\mathbf{V}_3$). Two internal inverse models of the arm, defined as neural mechanisms that can mimic the inverse of the input/output characteristics of the motor apparatus (Wolpert et al., 1998; Kawato, 1999), are learnt. These two inverse models, $\mathbf{V}_1$ at the cerebral level and $\mathbf{V}_2$ at the cerebellar level, calculate the necessary feedforward motor commands from the desired end-position, and cooperate during a global learning process.

The linear formalism required for an analytical solution is only valid locally, since the Jacobian **J** depends on the arm position. There are several methods that allow to overcome the problem of nonlinearity. One of them is the division of the working space in subspaces where the transformation is linear (Frolov, Rizek, 1995). Another plausible principle is to emphasize the proprioceptive information on arm position in the adaptation process and to learn a reorganization of this proprioceptive input through local activity-dependent synaptic adaptations, before combining it with visual information and calculating the motor commands (Baraduc et al., 2001). The use of linear operators in the solution of redundant inverse kinematics is prone to erratic behaviors (Klein and Huang 1983) and it is not proved that the method proposed here is immune from the same drawbacks, for example when the working point moves on a closed path. Nevertheless, we believe that the general idea of cerebral supervision of the cerebellar plasticity is more general and should be validated in a non-linear formalism.

Three "indeterminacy problems" of arm movement planning have been described: the infinite number of paths, the redundant degrees of freedom and the redundant muscles (Kawato et al., 1990), leading to the idea that the brain solves these ill-posed problems by some optimizing principles of optimal control, such as minimum-jerk (Flash and Hogan, 1985), minimum-torque-change (Uno et al., 1989), minimum-muscle-tension-change or minimum-commanded-torque-change (Nakano et al., 1999). The first indeterminacy problem of path selection is solved here by the equilibrium point theory, the WP path being determined by the straight line path of its equilibrium position. The last two indeterminacy problems are solved by the structural properties of the model (architecture and learning rules), and then are not ill-posed for the neural network. These properties provide learning of visuo-motor transformation to be pseudo-inverse to motor-visual transformation.

The model first shows that the unsupervised cerebral learning is faster than the supervised cerebellar learning and confirms that the global learning involves several phases (Ito, 1984). The choice of parameters determines the relative speeds of the cerebral and cerebellar learning processes, but the succession of their overlapping phases emerges from the structural model properties.

Secondly, the model explains how the cerebral cortex may determine the cerebellar climbing signals. These cerebral-origin olivary/climbing signals are not computed as explicit performance error (De Zeeuw et al., 1998). In the model, they are calculated in advance as a response to target presentation. Therefore, despite the low velocity of the olivo-cerebellar pathway and the low-frequency climbing fibre discharge, the timing of climbing and parallel fibre discharge is appropriate for optimal learning. The cortical origin of some climbing signals was experimentally established (Mano et al., 1986; Kitazawa et al., 1998) and was shown to originate from the lower layer V of the cerebral cortex (Saint-Cyr and Courville, 1980). Despite only cerebellar microzones receiving cerebral signals via the inferior olive were considered here, other microzones receiving peripheral signals may be added to the model, their output being added to signal $\mathbf{B}_2$.

The model also explains the time course of climbing fibre activity that was observed during motor learning tasks. Gilbert and Thach (1977) have shown that when a new perturbation is repeatedly applied to awake monkeys maintaining a forearm posture, a transient increase of climbing fibre activities (over several tens of trials) is observed, in relation to a long lasting decrease of the Purkinje cells simple spike activities. This transient increase corresponds to the transient increase of the efficiency of matrix $\mathbf{V}_1$ whose output projects to the inferior olive. The efficiency of this matrix increases from zero to a maximum at the first stage of learning and then decreases to zero, due to an increase of the cerebellar contribution to the movement performance.

The cerebellum not only controls simple movements, but also serial movements (Inhoff et al., 1989), and is known to participate to mental skills (Leiner et al., 1986). Further studies may use the present formalism to model the interactions of the lateral cerebellum with the premotor and prefrontal cortex (Dufossé et al., 1997).

The present formulation shows that: 1) any linear combination of cerebral motor commands may generate olivary signals able to drive the cerebellar learning process, 2) the climbing fibre activity supervising the cerebellar learning may originate from the generation of the cerebral commands, arising early enough to improve or even to replace these commands, before any error of performance would occur.

Acknowledgements

This work was supported by the Russian RFBR project 04-04-48989 and by CNRS position to author M.D.

References

Albus J.A. (1971) A theory of cerebellar function. Math. Biosci. 10: 25–61.

Arbib M.A., Érdi P., Szentagothai J. (1988) Neural organization. A Bradford book, Cambridge MA, MIT press.

Baraduc P., Guigon E., Burnod Y. (2001) Recoding arm position to learn visuomotor transformations. Cerebral Cortex 11(10): 906–917.

Baranyi A., Feher O. (1978) Conditioned changes of synaptic transmission in the motor cortex of the cat. Exp. Brain Res. 33: 283–298.

Burnod Y. (1989) An adaptive neural networks: the cerebral cortex, Masson, Paris.

Crépel F., Jaillard D. (1991) Pairing of pre- and postsynaptic activities in cerebellar Purkinje cells induce long-term changes in synaptic efficacy. J. Physiol. (London) 432: 123–141.

Daniel H., Levenes C., Crépel F. (1998) Cellular mechanisms of cerebellar LTD. Trends in Neurosci. 21: 401–407.

De Zeeuw C.I., Simpson J.I., Hoogenraad C.C., Galjart N., Koekkoek S.K.E., Ruigrok T.J.H. (1998) Microcircuitry and function of the inferior olive. Trends in Neurosci. 21: 391–400.

Dufossé M., Ito M., Jastreboff P. J., Miyashita Y. (1978) A neuronal correlate in rabbit's cerebellum to adaptive modification of the vestibulo-ocular reflex. Brain Res. 150: 611–616.

Dufossé M., Kaladjian A., Grandguillaume P. (1997) Origin of error signals during cerebellar learning of motor sequences. Behav. Brain Sci. 20(2): 249–250.

Feldman A.G. (1966) Functional tuning of the nervous structures during control of movement or maintenance of a steady posture: II. Controllable parameters of the muscle. Biophysics 11: 565–578.

Flash T., Hogan N. (1985) The coordination of arm movements: An experimentally confirmed mathematical model. J. Neurosci. 5(7): 1688–1703.

Frolov A.A., Dufossé M., Rizek S., Kaladjian A. (2000) On the possibility of linear modelling the human neuromuscular apparatus. Biol. Cybern. 82 (6): 499–515.

Frolov A.A., Řízek S. (1995) Model of neurocontrol of redundant systems. J. Comput. Appl. Math. 63: 465–473.

Georgopoulos A.P., Caminiti R., Kalaska J.F. (1984) Static spatial effects in motor cortex and area 5: quantitative relations in a two-dimensional space. Exp. Brain Res. 54: 447–454.

Gilbert P.F.C., Thach W.T. (1977) Purkinje cell activity during motor learning. Brain Res. 128: 309–328.

Gribble P.L., Ostry D.J., Sanguineti V., Laboissière R. (1998) Are complex control signals required for human arm movement? J Neurophysiol 79:1409–1424.

Golub G.H., Van Loan C.F. (1996) Matrix compputations. The John Hopkins University Press. Baltimore and London.

Inhoff A.W., Diener H.C., Rafal R.D., Ivry R. (1989) The role of cerebellar structures in the execution of serial movements. Brain 112: 565–581.

Iriki A., Pavlides C., Keller A., Asanuma H. (1989) Long-term potentation in the motor cortex. Science 245: 1385–1387.

Ito M. (1984) The cerebellum and neural control. Raven Press. New York.

Ito M., Sakurai M., Tongroach P. (1982) Climbing fibre induce depression of both mossy fibre responsiveness and glutamate sensitivity of cerebellar Purkinje cells. J. Physiol. (Lond.) 324: 113–134.

Jordan M.I., Rumelhart D.E. (1992) Forward models: supervised learning with a distal teacher. Cognitive Sci., 16: 307–354.

Kawato M. (1999) Internal models and trajectory planning. Curr. Op. Neurobiol. 9(6): 718–727.

Kawato M., Furukawa K., Suzuki R. (1987) A hierarchical neural-network model for control and learning of voluntary movement. Biol. Cybern. 57: 169–185.

Kawato M., Maeda Y., Uno Y., Suzuki R. (1990) Trajectory formation of arm movement by cascade neural network model based on minimum torque-change criterion. Biol. Cybern. 62: 275–288.

Kitazawa S., Kimura T., Yin P.B. (1988) Cerebellar complex spikes encode both destinations and errors in arm movements. Nature 392: 494–497.

Klein C., Huang C.H. (1983) Review of pseudo-inverse control for kinematically redundant manipulators. IEEE Trans. SMC-13: 245–250.

Kuperstein M. (1988) Neural model for adaptive hand-eye coordination for single postures. Science, 239:1308–1311.

Leiner H.C., Leiner A.L., Dow R.S. (1986) Does the cerebellum contribute to mental skills ? Behav. Neurosci. 100: 443–454.

Mano N.L., Kanazawa I., Yamamoto K.I. (1986) Complex-spike activity of cellular Purkinje cells related to wrist tracking movement in monkey. J. Neurophysiol 56: 137–158

Marr D. (1969) A theory of cerebellar cortex. J. Physiol. (London) 202: 437–470.

Meftah E.M., Rispal-Padel L. (1992) Synaptic plasticity in the thalamo-cortical pathway as one of the neurophysiological correlates of forelimb flexion conditioning: electrophysiological investication in the cat. J. Neurophysiol. 68: 908–926.

Nakano E., Imamizu H., Osu R., Uno Y., Gomi H., Yoshioka T., Kawato M. (1999) Quantitative examinations of internal representations for arm trajectory planning: minimum commanded torque change model. J. Neurophysiol. 81: 2140–2155.

Oyama E., MacDorman K.F., Agah A., Maedo T., Tachi S. (2001) Coordination transformation learning of hand position feedback controller with time delay. Neurocomputing, 38–40: 1503–1509.

Pananceau M., Rispal-Padel L., Meftah E.M. (1996) Synaptic plasticity in the interpositorubral pathway functionally related to forelimb flexion movements. J. Neurophysiol. 75: 2542–2561.

Saint-Cyr J.A., Courville J. (1980) Projections from the motor cortex, midbrain and vestibular nuclei to the inferior olive in tha cat: anatomical organization and functional correlates. In: The inferior olivary nucleus (pp 97–124). J. Courville, C. de Montigny and Y. Lamarre (Eds). Raven press, New York.

Schweighofer N., Arbib M.A., Kawato M. (1998) Role of cerebellum in reaching movements in humans. II. A neural model of the intermediate cerebellum. Europ. J. Neurosci. 10: 95–105.

Uno Y., Kawato M., Suzuki R. (1989) Formation and control of optimal trajectory in human multijoint arm movement. Biol. Cybern. 61: 89–101.

Wolpert D.M., Miall R.C., Kawato M. (1998) Internal models in the cerebellum. Trends in Cogn. Sci. 2(9): 338–347.

11. MOTOR PERFORMANCE AND REGIONAL BRAIN METABOLISM OF FOUR SPONTANEOUS MURINE MUTATIONS WITH DEGENERATION OF THE CEREBELLAR CORTEX

Robert Lalonde

Université de Rouen, Faculté de Médecine et de Pharmacie, INSERM U614, 76183 Rouen, Cedex France and CHUM/St-Luc, Neuroscience Research Center, Montreal, Canada

Catherine Strazielle

Université Henri Poincaré, Nancy I, Laboratoire de Pathologie Moléculaire et Cellulaire en Nutrition, INSERM U724, and Service de Microscopie Electronique, Faculté de Médecine, 54500 Vandoeuvre-les-Nancy France

Abstract

Four spontaneous mutations with cerebellar atrophy exhibit ataxia and deficits in motor coordination tasks requiring balance and equilibrium. These mutants were compared to their respective controls for regional brain metabolism assessed by histochemical staining of the mitochondrial enzyme, cytochrome oxidase (CO). The enzymatic activity of $Grid2^{Lc}$, $Grid2^{ho}$, $Rora^{sg}$, and $Reln^{rl}$ mutants was altered in cerebellum and cerebellar-related pathways at brainstem, midbrain, and telencephalic levels. The CO activity changes in cerebellar cortex and deep cerebellar nuclei as well as some cerebellar-related regions were linearly correlated with motor performance in stationary beam and rotorod tasks of $Grid2^{Lc}$, $Rora^{sg}$, and $Reln^{rl}$ mutants. These results indicate that in addition to its relation to neural activity, CO staining can be used as a predictor of motor capacity.

Keywords: cerebellum, motor control, equilibrium, cytochrome oxidase

1. *Introduction*

Spontaneous murine mutations with developmental defects causing degeneration of the cerebellar cortex have been known for many years (Lalonde and Strazielle 1999). But only recently have genes been identified, namely $Grid2^{Lc}$ (*Lurcher*), $Grid2^{ho}$ (*hotfoot*), $Rora^{sg}$ (*staggerer*), and $Reln^{rl}$ (*reeler*). The cerebellar mutants exhibit cerebellar ataxia (wide-spread gait) and motor coordination deficits in tasks requiring balance and equilibrium.

2. *Neuropathology in Cerebellar Mutant Mice*

The neuropathology of the semi-dominant *Lurcher* (allele symbol: *Lc*) mutation ($Grid2^{Lc}$) is caused by a gain-in-malfunction of *Grid2* located on chromosome 6. This gene encodes an ionotropic glutamate receptor (GluRδ2) functionally related with AMPA receptors (Landsend et al. 1997) and predominantly expressed in cerebellar Purkinje cells (Zuo et al. 1997).

While homozygous (*Grid2*^{Lc}/*Grid2*^{Lc}) mutants cannot survive beyond the first postnatal day because of defective suckling caused by brainstem damage (Cheng and Heintz 1997; Resibois et al. 1997), heterozygous (*Grid2*^{Lc}/-) mutants have been tested for motor control and shown to be deficient during developmental (Thullier et al. 1997) and adult (Lalonde and Strazielle 1999) periods. The increased permeability of the mutated GluRδ2 channel to calcium (Wollmuth et al. 2000) may be responsible for the nearly complete degeneration of Purkinje cells occurring from the second to the fourth postnatal week (Caddy and Biscoe 1979). The massive degeneration of granule cells is attributed to the loss of the trophic influence exerted by Purkinje cells (Vogel et al. 1991). In a similar fashion, the 60 to 75% decrease in the number of inferior olive cells (Caddy and Biscoe 1979; Heckroth and Eisenman 1991) and the 30% decrease in the number of deep cerebellar nuclei (Heckroth 1994) appear to be secondary consequences of Purkinje cell atrophy.

Two recessive *hot-foot* (allele symbol: *ho*) mutations (4-J and Nancy) cause different deletions of the coding sequences of the *Grid2* gene (Lalouette et al. 1998, 2001). At least for the *4-J* allele, the truncated GluRδ2 protein was expressed in the soma of Purkinje cells but without transport to the cell surface (Matsuda and Yuzaki 2002). In an opposite manner to *Grid2*^{Lc}, the encoded protein appears non-functional, as the neuropathological and behavioral phenotypes of *Grid2*^{ho} mutants were similar to those of targeted *Grid2* null mutants (Kashiwabuchi et al. 1995). The *Grid2*^{ho} mutants are characterized by defective innervation of Purkinje cells by parallel fibers and by a mild loss of cerebellar granule cells (Guastavino et al. 1990).

The *Grid2*^{ho} model has been bred with *Grid2*^{Lc} to obtain the double *Grid2*$^{ho/Lc}$ mutant (Selimi et al. 2003). The type of cerebellar atrophy seen in *Grid2*$^{ho/Lc}$ double mutants is more similar to *Grid2*$^{Lc/+}$ than *Grid2*$^{ho/ho}$, but during development, Purkinje cell number was lower in the double mutant than in the single *Grid2*$^{Lc/+}$ mutant.

The recessive *Rora*^{sg} mutation causes a deletion of the *Rora* gene situated on chromosome 9, encoding the retinoid-like nuclear receptor involved in neuronal differentiation and maturation, particularly expressed in Purkinje cells (Hamilton et al. 1996; Nakagawa et al. 1997). The retinoid-like protein appears non-functional, as the neuropathological and behavioral phenotypes of *Rora*^{sg} were similar to those of *Rora* null mutants (Steinmayr et al. 1998). The Purkinje cell number of *Rora*^{sg} mutants declined on embryonic day 14 and reached 25% of normal values at the end of the first postnatal month (Herrup and Mullen 1979). Thus, the Purkinje cell loss begins at an earlier stage of development but is less complete than *Grid2*^{Lc} (Caddy and Biscoe 1979). The remaining Purkinje cells in *Rora*^{sg} mutants were reduced in size, ectopically positioned, and lacked the tertiary dendritic spines receiving synaptic contacts from parallel fibers (Sotelo 1975). The secondary degeneration of granule cells occurred soon after their migration (Herrup 1983) and was nearly complete by the end of the first postnatal month (Landis and Sidman, 1978). Unlike *Grid2*^{Lc} and *Grid2*^{ho} mutants, the massive degeneration of the cerebellar cortex makes the molecular and granule cell layers difficult to distinguish. Although deep cerebellar nuclei were present in normal numbers, their volume was reduced in *Rora*^{sg} mutants (Roffler-Tarlov and Herrup 1981). Presumably because of Purkinje cell loss, the number of inferior olive neurons decreased by 60% on postnatal day 24 (Shojaeian et al. 1985a) and remained lower than normal in adults (Blatt and Eisenman 1985a).

The autosomal recessive *Reln*^{rl} mutation causes a disruption of the *Reln* gene, located on chromosome 5 (Beckers et al. 1994; D'Arcangelo et al. 1995). This gene encodes an extracellular matrix protein involved in neural adhesion and migration at critical stages of development (Beckers et al. 1994; D'Arcangelo et al. 1995, 1999; Hack et al. 2002; Trommsdorff et al. 1999). The *Reln*^{rl} mutant displays abnormal architectonic organization and cell ectopias, but with preserved anatomical connections in cerebellum, inferior olive, hippocampus, and neocortex (Mariani et al. 1977; Stanfield and Cowan 1979; Goffinet 1983; Blatt and Eisenman 1988; Heckroth et al. 1989; Terashima et al. 1986). The principal cerebellar cell type depleted by the mutation is the granule cell population. The Purkinje cell loss reached approximately 50%, the remaining Purkinje cells being malpositionned and grouped in a central mass (Heckroth et al. 1989). Presumably as a consequence of the Purkinje cell deficit, the number of inferior olive cells diminished by 20% (Blatt and Eisenman 1985b; Shojaeian et al. 1985b). Despite ectopic positioning, the zonal pattern of climbing fiber projections was maintained (Blatt and Eisenman 1988), but with Purkinje cells abnormally innervated by more than one climbing fiber, as found with other dysgranular cerebellar mutants such as *Rora*^{sg} (Mariani 1982; Mariani et al. 1977).

3. *Motor Coordination Deficits of Cerebellar Mutant Mice*

The main measure used for testing motor coordination in mice is the time elapsed before falling from

TABLE 1. Regional brain variations of cytochrome oxidase activity in cerebellum of *Grid2*^{Lc} (*Lc*), *Grid2*^{ho} (*ho*), *Rora*^{sg} (*sg*), and *Reln*^{rl} (rl) mutants

Brain region	*Lc*	*ho*	*sg*	*rl*
Cortex				
-molecular	nc	↑	nc	nc
-Purkinje	depl	nc	depl	↑
-granular	nc	nc	depl	nc
Deep nuclei				
-fastigial	↑	nc	↑	nc
-interpositus	↑	nc	↑	↓*
-dentate	↑	nc	↑	↓*

↓ decreased, ↑ increased, nc not changed, depl = too severily depleted or not dissociable from other layers, *measured as a single region designated "roof nuclei"

TABLE 2. Variations of cytochrome oxidase activity in cerebellar-related pathways of *Grid2*^{Lc} (*Lc*), *Grid2*^{ho} (*ho*), *Rora*^{sg} (*sg*), and *Reln*^{rl} (*rl*) mutants

Brain regions	*Lc*	*ho*	*sg*	*rl*
Neocortex				
-primary motor	nc	nc	nc	↑↓*
-eye field	nc	↑	nc	nc
Thalamus				
-ventrolateral	↑	↑	nc	nc
-ventromedial	nc	nc	nc	nc
-dorsomedial	nc	nc	↓	nc
-lateral geniculate	nc	nc	↓	↑
-midline	nc	↑	↓	nc
Red nucleus	↑	nc	↑	nc
Interpeduncular	↑	nc	↑	nc
Dorsal raphe	↑	nc	nc	nc
Vestibular nuclei				
-medial	nc	nc	↑	nc
-lateral	↑	nc	↑	nc
Pontine nuclei				
-medial	nc	nc	nc	↑
-lateral	nc	nc	nc	nc
Inferior olive	↓	nc	nc	nc

↓ decreased, ↑ increased, nc not changed, *dependent on cell layer

a narrow surface. In the stationary beam test, mice move along a narrow rod and the distance travelled can be used as an auxiliary measure (Lalonde and Strazielle 1999). In the rotorod test, mice are placed on a beam revolving around its longitudinal axis, so that synchronized forward locomotion is neccessary in order to avoid a fall. In the suspended wire test, mice are placed upside-down on a thin horizontal wire. The coat-hanger is a variation of this standard test and provides the opportunity of estimating movement time, as latencies before the suspended mice reach the extremity of the horizontal wire and begin to climb on one of the diagonal bars of the triangular-shaped apparatus are measured.

By comparison to non-ataxic littermates controlled for age and sex, latencies before falling of *Grid2*^{Lc} mutants decreased in stationary beam, coat-hanger, and rotorod tests (Caston et al. 1995; Lalonde et al. 1992, 1995, 1996). The same deficits were found in *Grid2*^{ho} (Krémarik et al. 1998; Lalonde et al. 1995,

1996; Lalouette et al. 2001), $Rora^{sg}$ (Deiss et al. 2000; Lalonde et al. 1995, 1996) and $Reln^{rl}$ (Lalonde et al. 2003a) mutants, attesting to the sensitivity of these measures to cerebellar dysfunction. The motor deficits observed in $Grid2^{ho}$ mice were potentiated after interbreeding with $Grid2^{Lc}$ (Lalonde et al. 2003).

4. *Regional Brain Metabolism in Cerebellar Mutant Mice*

The effects of chronic cerebellar lesions seen in $Grid2^{Lc}$, $Grid2^{ho}$, $Rora^{sg}$, and $Reln^{rl}$ mutants on regional brain metabolism were estimated by cytochrome oxidase (CO) staining (Tables 1 and 2). CO is the fourth enzyme of the mitochondrial electron transfer chain responsible for oxidative phosphorylation and the production of adenosine triphosphate (ATP), an essential source of cellular energy. Unlike glucose uptake, CO labelling is a specific marker of neuronal activity, as glial cells contribute only minimally to oxidative metabolism (Wong-Riley 1989; Gonzalez-Lima and Jones 1994). Because CO histochemistry is expressed as a function of tissue weight, a loss in neuronal numbers does not necessarily lead to a reduction in enzymatic activity, as the metabolic level of remaining neurons may be normal. The metabolic level of an atrophied region may even increase, as brain tissue loss may be disproportionally large relative to the high metabolic level of remaining neurons.

4.1. CEREBELLUM

As presented in Table 1, CO activity was altered in cerebellar subregions depending to the nature of the lesion and its developmental period.

CO activity was first examined in $Grid2^{Lc}$ mutant mice (Strazielle et al. 1998). Despite massive degeneration of the cerebellar cortex, CO activity in shrunken but still identifiable molecular and granule layers was unchanged in $Grid2^{Lc}$ mutants (Fig. 1b and 2b). However, their CO activity was higher than that of controls in deep cerebellar nuclei (Fig. 2b), which receive GABAergic afferent impulses from Purkinje cells. A plausible reason for this hypermetabolism is the lost inhibitory input due to depleted Purkinje cells, as the metabolic level required for excitatory synapses is higher than that of inhibitory ones (Wong-Riley 1989). A similar pattern was revealed in $Rora^{sg}$ mutants, although unlike $Grid2^{Lc}$, their molecular layer cannot easily be distinguished from the granular layer (Fig. 1d). Like the $Grid2^{Lc}$ model, severe cerebellar cortical degeneration in $Rora^{sg}$ mutants did not change CO activity, indicating no ongoing hypometabolism of remaining neurons (Fig. 2d). Again like the $Grid2^{Lc}$ model, CO activity in $Rora^{sg}$ deep cerebellar nuclei was elevated (Fig. 2d), probably because of lost GABAergic inhibitory input from depleted Purkinje cells or because of increased excitatory input from afferent (climbing or mossy) fibers.

Unlike previous mutants, the metabolic activity of the cerebellar cortex was modified in $Grid2^{ho}$ mutant mice (Krémarik et al. 1998). Indeed, CO activity in molecular layer was higher in $Grid2^{ho}$ mutants than their respective controls, perhaps due to upregulated activity of cerebellar afferents in response to defective dendritic organization of Purkinje cells (Guastavino et al. 1990). A second difference from previous mutants is the unchanged CO activity seen in $Grid2^{ho}$ deep cerebellar nuclei (Fig. 2c), probably because Purkinje cells are miminally depleted or perhaps not at all (Fig. 1c).

In unfoliated $Reln^{rl}$ cerebellar cortex, molecular and granule cell layers are still identifiable (Fig. 1e), as cells are widely but not randomly scattered (Strazielle et al. 2005). Except for a few correctly positioned cells, most Purkinje cells are not located in their regular single monolayer between the higher molecular layer and the lower granular layer, but instead are dispersed throughout cerebellar cortex (Fig. 2e). It was not possible to distinguish the interpositus from the dentate, and therefore a single measure was obtained, designated as roof nuclei. Like $Grid2^{Lc}$ cerebellar cortex, CO activity was unchanged in molecular and granule cell layers of $Reln^{rl}$ mutants. However, CO activity increased in correctly positioned Purkinje cells and diminished in roof nuclei (Fig. 2e). Quantitative optical density readings of sections stained with methylene blue demonstrated no significant change of coloration per surface unit or per total surface area in roof nuclei of $Reln^{rl}$ mutants, indicating that higher CO activity was not the consequence of neuronal atrophy. Instead, decreased metabolic level of roof nuclei may be due to the influence of hypermetabolic Purkinje cells or else to ongoing degenerative processes.

4.2. CEREBELLAR-RELATED REGIONS

In concordance with the hypothesis of metabolic consequences resulting from missing Purkinje cells, CO activity increased not only in deep cerebellar nuclei of $Grid2^{Lc}$ mutants but also in lateral vestibular nucleus, structures receiving direct Purkinje cell input (Strazielle et al. 1998). As presented in Table 2, CO activity was elevated as well in lateral and medial vestibular nuclei of $Rora^{sg}$ mutants (Deiss et al. 2000). On the contrary, no such effect was observed in $Grid2^{ho}$ (Krémarik et al. 1998) and $Reln^{rl}$ (Strazielle et al. 2005) mutants, characterized by milder Purkinje cell losses.

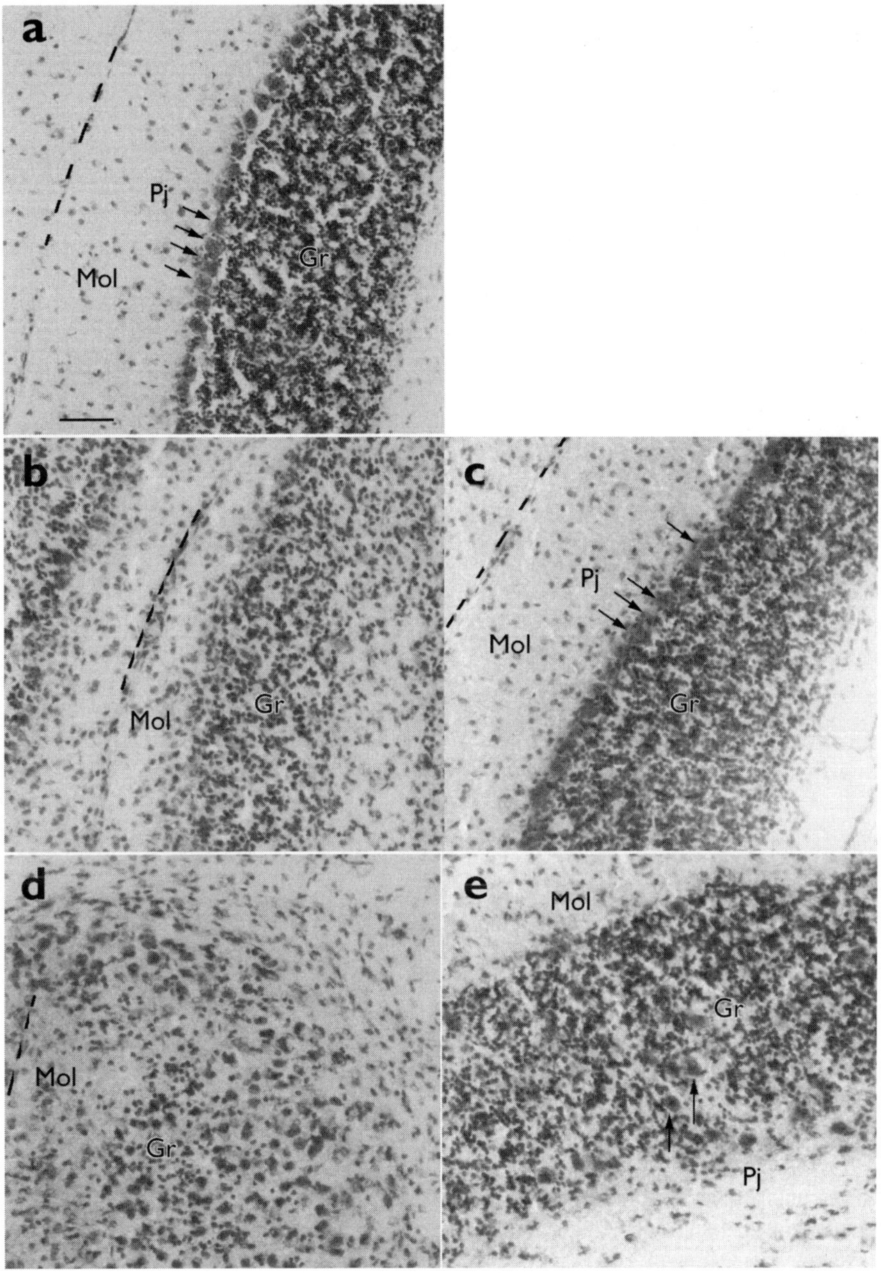

FIGURE 1. Methylene blue staining of cerebellar cortex (lobule simplex) in control (a), *Grid2^{Lc}* (b), *Grid2^{ho}* (c), *Rora^{sg}* (d), and *Reln^{rl}* (e) mutant mice at the −1.72 posterior plane (Scale bar, 150 μm). The three cortical layers of cerebellar cortex are easily distinghishable in the control section (a). In the *Grid2^{Lc}* (b) mutant, note the severe lobule atrophy, particularly evident in molecular layer, as well as the weak cell density of granular layer. Purkinje cells have totally disappeared. In the *Grid2^{ho}* (c) mutant, the cerebellar cortex is very similar to control, except for mild atrophy of the molecular layer. The cerebellar cortex of the *Rora^{sg}* (d) mutant has lost its laminar organization. A few ectopic Purkinje cells with smaller size than those of controls are present in an undefined layer. In unfoliated *Reln^{rl}* (e) cerebellar cortex, molecular and granule cell layers were still identifiable. Except for a few correctly positioned cells, most Purkinje cells were no longer situated in a single monolayer, but instead scattered in lower parts of cerebellar cortex. Gr = granular layer, Mol = molecular layer, Pj = Purkinje cell.

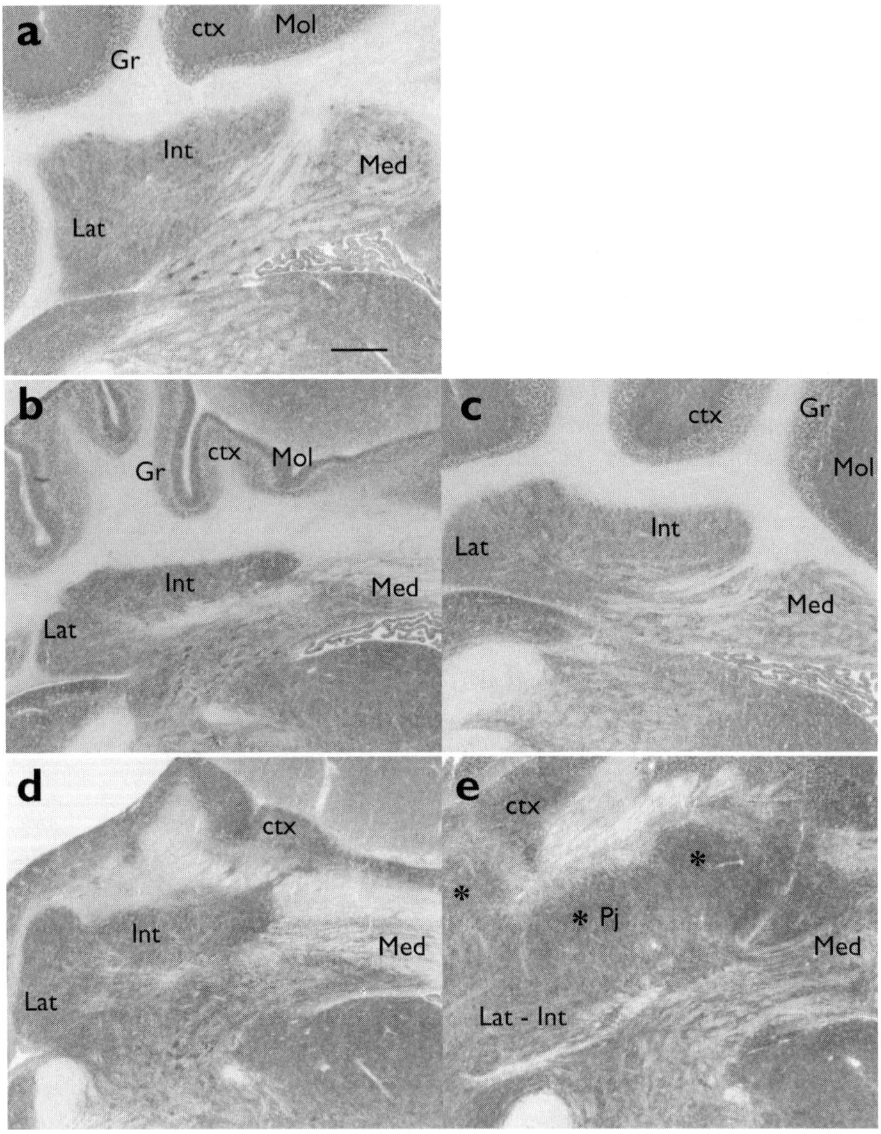

FIGURE 2. Cytochrome oxidase (CO) labelling of cerebellar region analyzed at the −2.08 mm posterior plane in *Grid2*^{Lc} (b), *Grid2*^{ho} (c), *Rora*^{sg} (d), and *Reln*^{rl} (e) mutants in comparison with same region of a control mouse (a). (Scale bar, 250 μm). In the *Grid2*^{Lc} mutant (b), CO staining intensity significantly increased in deep cerebellar nuclei and in lateral vestibular nucleus. Note the preservation of CO labelling in cerebellar cortex despite atrophy of the structure. The CO labelling pattern of *Grid2*^{ho} cerebellum (c) is very similar to control (a) except for a mild increase in molecular layer of cerebellar cortex. In the *Rora*^{sg} mutant (d), note the more intense labelling of cerebellar deep nuclei and preservation of CO labelling in cerebellar cortex despite complete dysorganization. In the *Reln*^{rl} (e) mutant, the tri-lamination of the cortex is preserved in the superficial portion of cerebellar cortex, where CO activity remained unchanged in molecular and granule cell layers. Under the central masses of ectopic Purkinje cells, roof nuclei composed of interpositus (Int), and dentate or lateral (Lat) nuclei, presented lower CO density of labelling when compared with control, whereas CO labelling of well-defined fastigial or medial (Med) nucleus, remained unchanged. ctx = cerebellar cortex, Gr = granular layer, Int = interpositus nucleus, Lat = lateral cerebellar deep nucleus (dentate nucleus), Med = medial cerebellar deep nucleus (fastigial nucleus), Mol = molecular nucleus.

The CO activity of *Grid2^{Lc}* mutants was elevated in additional afferent/efferent regions directly connected with the cerebellum, such as magnocellular red nucleus, interpeduncular nucleus, dorsal raphe, and ventrolateral thalamus, possibly as a result of hypermetabolic deep cerebellar nuclei, which send excitatory impulses at least to the red nucleus. By contrast, CO activity decreased in inferior olive, a brain region that undergoes atrophy in this mutant (Caddy and Biscoe 1979; Heckroth and Eisenman 1991), probably through retrograde degeneration secondary to the Purkinje cell loss. CO histochemistry may be considered as the metabolic signature of the ongoing degenerative process, with hypometabolism eventually leading to cell death. The absence of hypometabolism in cerebellar cortex probably indicates that in the adult mutant the degenerative process has reached a plateau.

As with *Grid2^{Lc}*, CO activity was altered in some cerebellar-related regions in *Grid2^{ho}* (Krémarik et al. 1998), *Rora^{sg}* (Deiss et al. 2000), and *Reln^{rl}* (Strazielle et al. 2005) mutants. The augmented CO activity found in *Grid2^{Lc}* ventrolateral thalamus was matched in *Grid2^{ho}*, but not at the level of red, interpeduncular, and dorsal raphe nuclei. In further contrast, CO activity was unchanged in inferior olive.

The CO staining pattern of *Rora^{sg}* resembles that of *Grid2^{Lc}* mutants in terms of elevated activity in red and interpeduncular nuclei, but differ in respect to unchanged activity in dorsal raphe and inferior olive. CO staining in the *Rora^{sg}* inferior olive was normal despite cellular atrophy (Blatt and Eisenman 1985a; Shojaeian et al. 1985), presumably because ongoing degenerative processes had stopped at the time when CO measures were taken (one year of age). In contrast to both *Grid2^{Lc}* and *Grid2^{ho}* mutants, CO activity was unchanged in ventrolateral but decreased in other thalamic subregions receiving cerebellar input, namely ventroanterior, dorsomedial, lateral geniculate, and posterior nuclei, and also in thalamic subregions without such input, such as midline nuclei. This decreased metabolic activity may be a secondary consequence of cerebellar damage, but more likely a direct result of *Rora* deletion, a gene highly expressed in this region (Sashihara et al. 1996). As reported in *Grid2^{Lc}* and *Grid2^{ho}* mice, CO activity was unchanged in *Rora^{sg}* medial and lateral pontine nuclei, one source of mossy fiber afferent input to cerebellum. Moreover, CO staining was normal in motor cortex of all three mutants, reflecting the relative absence of transsynaptic changes in regions higher than the diencephalon.

In *Reln^{rl}* mutants, CO activity increased in only one thalamic subregion receiving cerebellar afferents: lateral geniculate, and not in ventrolateral, ventroanterior, dorsomedial, and posterior nuclei (Strazielle et al. 2005). Unlike other mutants, CO activity was not altered in red, interpeduncular, dorsal raphe, and inferior olive, but was lower than controls in medial pontine nuclei (Strazielle et al. 2005). In view of cell ectopias existing in neocortex, CO changes were more prominent at this level than the other mutants. For example, in *Reln^{rl}* primary motor cortex, CO activity was lower in granule cell layer (M1gr) but higher in polymorphic (M1poly) cell layer. CO activity was also higher in polymorphic cell layer of primary somatosensory (S1poly) and piriform cortices.

5. *Brain-Behavior Relations in Cerebellar Mutant Mice*

Linear regressions were undertaken in cerebellar mutants for the purpose of determining whether specific brain regions are associated with motor deficits. In *Grid2^{Lc}* mutants, latencies before falling from the rotorod were positively correlated with abnormally high CO activity in magnocellular red nucleus (Strazielle et al. 1998). These results indicate that augmented CO activity is related to improved performances. Because the rubrospinal tract discharges in phase with the locomotor cycle, it may be hypothesized that the red nucleus takes over from a dysfunctional cerebellum. No such relation was found in regard to elevated metabolism in ventrolateral thalamus of *Grid2^{Lc}* and *Grid2^{ho}* mutants (Krémarik et al. 1998; Strazielle et al. 1998). However, high CO activity in *Rora^{sg}* medial vestibular nucleus was associated with longer distances travelled on the stationary beam (Deiss et al. 2000). In contrast, high CO activity in either interpositus or dentate nuclei was linearly correlated with shorter distances travelled on the stationary beam and poorer rotorod performances.

Some linear correlations between stationary beam performances and areas showing either increased (Purkinje cell and S1poly) or diminished (roof nuclei and M1gr) CO activity were significant in *Reln^{rl}* mutants as well (Lalonde et al. 2005). Elevated CO activity in Purkinje cells was associated with poorer stationary beam performance. This result resembles the association existing between poor stationary beam and rotorod performances and elevated deep nuclei enzymatic activity in *Rora^{sg}* mutants. However, the results were more variable in respect to the roof nuclei. Indeed, CO activity was linearly correlated with poorer performances of *Reln^{rl}* mice on a small stationary beam but with better performances on a larger one.

High CO activity in S1poly was also associated with better scores on the large beam. Low CO activity in M1gr was correlated with a longer distance travelled but with shorter latencies before falling from the large beam, indicating that those mutants with low CO activity were less immobile but with an increased risk of falling (Lalonde et al. 2004).

Overall, these data show that CO activity in cerebellum and related regions is a significant predictor of motor performances in cerebellar mutant mice. These results add depth to the already known relation between the activity of this enzyme and neural activity (Wong-Riley, 1989). Moreover, these results are congruent with significant linear correlations existing between the severity of motor symptoms on one hand and glucose utilization on the other in brain regions of patients with spinocerebellar ataxias (Kluin et al. 1988; Rosenthal et al. 1988).

References

Beckers MC, Bar I, Huynh-Thu T, Dernoncourt C, Brunialti AL, Montagutelli X, Guenet JL, Goffinet AM (1994) A high-resolution genetic map of mouse chromosome 5 encompassing the reeler (rl) locus. Genomics 23:685–690.

Blatt GJ, Eisenman LM (1985a) A qualitative and quantitative light microscopic study of the inferior olivary complex in the adult *staggerer* mutant mouse. J Neurogenet 2:51–66.

Blatt GJ, Eisenman LM (1985b) A qualitative and quantitative light microscopic study of the inferior olivary complex of normal, *reeler*, and *weaver* mutant mice. J Comp Neurol 232:117–128.

Blatt GJ, Eisenman LM (1988) Topographic and zonal organization of the olivocerebellar projection in the *reeler* mutant mouse. J Comp Neurol 267:603–615.

Caddy KWT, Biscoe TJ (1979) Structural and quantitative studies on the normal C3H and *Lurcher* mutant mouse. Philos Trans Roy Soc Lond (Biol) 287:167–201.

Caston J, Vasseur F, Stelz T, Chianale C, Delhaye-Bouchaud N, Mariani J (1995) Differential roles of cerebellar cortex and deep cerebellar nuclei in the learning of equilibrium behavior: studies in intact and cerebellectomized control and *Lurcher* mutant mice. Dev Brain Res 86: 311–316.

Cheng S-W, Heintz N (1997) Massive loss of mid- and hindbrain neurons during embryonic development of homozygous *Lurcher* mice. J Neurosci 17:2400–2407.

D'Arcangelo G, Homayouni R, Keshvara L, Rice DS, Sheldon M, Curran T (1999) Reelin is a ligand for lipoprotein receptors. Neuron 24:471–479.

D'Arcangelo G, Miao GG, Chen S-C, Soared HD, Morgan JI, Curran T (1995) A protein related to extracellular matrix proteins deleted in the mouse mutant *reeler*. Nature 374:719–723.

Deiss V, Strazielle C, Lalonde R (2000) Regional brain variations of cytochrome oxidase activity and motor coordination in *staggerer* mutant mice. Neuroscience 95:903–911.

Goffinet AM (1983) The embryonic development of the inferior olivary complex in normal and *reeler* mutant mice. J Comp Neurol 219:10–24.

Gonzalez-Lima F, Jones D (1994) Quantitative mapping of cytochrome oxidase activity in the central auditory system of the gerbil: a study with calibrated activity standards and metal-intensified histochemistry. Brain Res 660:34–49.

Guastavino J-M, Sotelo C, Damez-Kinselle I (1990) *Hotfoot* murine mutation: behavioral effects and neuroanatomical alterations. Brain Res 523:199–210.

Hack I, Bancila M, Loulier K, Carroll P, Cremer H (2002) Reelin is a detachment signal in tangential chain-migration during postnatal neruogenesis. Nature Neurosci 5:939–945.

Hamilton BA, Frankel WN, Kerrebrock AW, Hawkins TL, Fitzhugh W, Kusumi K, Russell LB, Mueller KL, Van Burkel V, Birren BW, Krugiyak L, Lander EE (1996) Disruption of the nuclear hormone receptor ROR in *staggerer* mice. Nature 379:736–739.

Heckroth JA (1994) Quantitative morphological analysis of the cerebellar nuclei in normal and *Lurcher* mutant mice. I. Morphology and cell number. J Comp Neurol 343:173–182.

Heckroth JA, Eisenman LM (1991) Olivary morphology and olivocerebellar atrophy in adult *Lurcher* mutant mice. J Comp Neurol 312:641–651.

Heckroth JA, Goldowitz D, Eisenman LM (1989) Purkinje cell reduction in the *reeler* mutant mouse: a quantitative immunohistochemical study. J Comp Neurol 279:546–555.

Herrup K (1983) Role of *staggerer* gene in determining cell number in cerebellar cortex. I. Granule cell death is an indirect consequence of *staggerer* gene action. Dev Brain Res 11:267–274.

Herrup K, Mullen RJ (1979) Regional variation and absence of large neurons in the cerebellum of the *staggerer* mouse. Brain Res 172:1–12.

Kashiwabuchi N, Ikeda K, Araki K, Hirano T, Shibuki K, Takayama C, Inoue Y, Kutsuwada T, Yagi T, Kang Y, Aizawa S, Mishina M (1995) Impairment of motor coordination, Purkinje cell synapse formation, and cerebellar long-term depression in GluR delta 2 mutant mice. Cell 81:245–252.

Kluin KK, Gilman S, Markel DS, Koeppe RA, Rosenthal G, Junck L (1988) Speech disorders in olivopontocerebellar atrophy correlate with positron emission tomography findings. Ann Neurol 23:547–554.

Krémarik P, Strazielle C, Lalonde R (1998) Regional brain variations of cytochrome oxidase activity and motor coordination in *hot-foot* mutant mice. Eur J Neurosci 10:2802–2809.

Lalonde R, Strazielle C (1999) Motor performance of spontaneous murine mutations with cerebellar atrophy. In: W Crusio, R Gerlai (Eds) Handbook of molecular-genetic techniques for brain and behavior research (Techniques in the Behavioral and Neural Sciences, vol 13). Amsterdam: Elsevier, pp 627–637.

Lalonde R, Bensoula AN, Filali M (1995) Rotorod sensorimotor learning in cerebellar mutant mice. Neurosci Res 22:423–426.

Lalonde R, Botez MI, Joyal CC, M. Caumartin M (1992) Motor deficits in *Lurcher* mutant mice. Physiol Behav 51:523–525.

Lalonde R, Filali M, Bensoula AN, Lestienne F (1996) Sensorimotor learning in three cerebellar mutant mice. Neurobiol Learn Mem 65:113–120.

Lalonde R, Hayzoun K, Derer M, Mariani J, Strazielle C (2004) Neurobehavioral evaluation of $Reln^{rl}$ mutant mice: correlations with cytochrome oxidase activity. Neurosci Res 49:297–305.

Lalonde R, Hayzoun K, Selimi F, Mariani J, Strazielle C (2003) Motor coordination in mice with *hot-foot*, *Lurcher*, and double mutations of the *Grid2* gene encoding the delta-2 excitatory amino acid receptor. Physiol Behav 80:333–339.

Lalouette A, Guénet J-L, Vriz S (1998) *Hot-foot* mutations affect the δ2 glutamate receptor gene and are allelic to *Lurcher*. Genomics 50:9–13.

Lalouette A, Lohof A, Sotelo C, Guénet J, Mariani J (2001) Neurobiological effects of a null mutation depend on genetic context: comparison between two *hotfoot* alleles of the delta-2 ionotropic glutamate receptor. Neuroscience 105:443–455.

Landis DMD, Sidman RL (1978) Electron microscopic analysis of postnatal histogenesis in the cerebellar cortex of *staggerer* mutant mice. J Comp Neurol 179:831–864.

Landsend AS, Amiry-Moghaddam M, Matsubara A, Bergersen L, Usami S, Wenthold RJ, Ottersen OP (1997) Differential localization of delta glutamate receptors in the rat cerebellum: coexpression with AMPA receptors in parallel fiber-spine synapses and absence from climbing fiber-spine synapses. J Neurosci 17:834–842.

Mariani J (1982) Extent of multiple innervation of Purkinje cells by climbing fibers in the olivocerebellar system of *weaver*, *reeler* and *staggerer* mutant mice. J Neurobiol 13:119–126.

Mariani J, Crepel F, Mikoshiba K, Changeux J-P, Sotelo C (1977) Anatomical, physiological and biochemical studies of the cerebellum from *reeler* mutant mouse. Phil Trans Roy Soc London (Biol) 281:1–28.

Matsuda S, Yuzaki M (2002) Mutation in *hotfoot-4J* mice results in retention of δ2 glutamate receptors in ER. Eur J Neurosci 16:1507–1516.

Nakagawa S, Watanabe M, Inoue Y (1997) Prominent expression of nuclear hormone receptor RORα in Purkinje cells from early development. Neurosci Res 28:177–184.

Resibois A, Cuvelier L, Goffinet AM (1997) Abnormalities in the cerebellum and brainstem in homozygous *Lurcher* mice. Neuroscience 80:175–190.

Roffler-Tarlov S, Herrup K (1981) Quantitative examination of the deep cerebellar nuclei in the *staggerer* mutant mouse. Brain Res 215:49–59.

Rosenthal G, Gilman S, Koeppe RA, Kluin KJ, Markel DS, Junck L, Gebarski SS (1988) Motor dysfunction in olivopontocerebellar atrophy is related to cerebral metabolic rate studies with positron emission tomography. Ann Neurol 24:414–419.

Sashihara S, Felts PA, Waxman SG, Matsui T (1996) Orphan nuclear receptor *ROR* gene: isoform-specific spatiotemporal expression during postnatal development of brain. Mol Brain Res 42:109–117.

Selimi F, Lohof AM, Heitz S, Lalouette A, Jarvis CI, Bailly Y, Mariani J (2003) Lurcher *GRID2*-induced death and depolarization can be dissociated in cerebellar Purkinje cells. Neuron 37:813–819.

Shojaeian H, Delhaye-Bouchaud N, Mariani J (1985a) Decreased number of cells in the inferior olivary nucleus of the developing *staggerer* mouse. Dev Brain Res 21:141–146.

Shojaeian H, Delhaye-Bouchaud N, Mariani J (1985b) Neuronal death and synapse elimination in the olivocerebellar system. II. Cell counts in the inferior olive of adult X-irradiated rats and *weaver* and *reeler* mice. J Comp Neurol 232:309–318.

Sotelo C (1975) Dendritic abnormalities of Purkinje cells in the cerebellum of neurologic mutant mice (*weaver* and *staggerer*). In: G.W. Kreutzberg (Ed.) Physiology and pathology of dendrite (Advances in Neurology). New York: Raven Press, pp 335–351.

Stanfield BB, Cowan WM (1979) The morphology of the hippocampus and dentate gyrus in normal and *reeler* mutant mice. J Comp Neurol 185:393–422.

Steinmayr M, André E, Conquet F, Rondi-Reig L, Delhaye-Bouchaud N, Auclair N, Daniel H, Crépel F, Mariani J, Sotelo C, Becker-André M (1998) *Staggerer* phenotype in retinoid-related orphan receptor α-deficient mice. Proc Natl Acad Sci USA 95:3960–3965.

Strazielle C, Krémarik P, Ghersi-Egea J-F, Lalonde R (1998) Regional brain variations of cytochrome oxidase activity and motor coordination in *Lurcher* mutant mice. Exp Brain Res 121:35–45.

Strazielle C, Hayzoun K, Derer M, Mariani J, Lalonde R (2005) Regional brain variations of cytochrome oxidase activity in *Reln*^{rl} mutant mice. Synapse (submitted).

Terashima T, Inoue K, Inoue Y, Yokoyama M, Mikoshiba K (1986) Observations on the cerebellum of normal-*reeler* mutant mouse chimera. J Comp Neurol 252:264–278.

Trommsdorff M, Gotthardt M, Hiesberger T, Shelton J, Stockinger W, Nimpf J, Hammer RE, Richardson JA, Herz J. (1999) Reeler/disabled-like disruption of neuronal migration in knockout mice lacking the VLDL receptor and ApoE receptor 2. Cell 97:689–701.

Thullier F, Lalonde R, Cousin X, Lestienne F (1997) Neurobehavioral evaluation of *Lurcher* mutant mice during ontogeny. Dev Brain Res 100:22–28.

Vogel MW, McInnes M, Zanjani HS, Herrup K (1991) Cerebellar Purkinje cells provide target support over a limited spatial range: evidence from *Lurcher* chimeric mice. Dev Brain Res 64:87–94.

Wollmuth L, Kuner T, Jatzke C, Seeburg PH, Heintz N, Zuo J (2000) The *Lurcher* mutation identifies δ2 as an AMPA/kainate receptor-like channel that is potentiated by Ca^{2+}. J Neurosci 20:5973–5980.

Wong-Riley MTT (1989) Cytochrome oxidase: an endogenous metabolic marker for neuronal activity. Trends Neurosci 12:94–101.

Zuo J, De Jager Pl, Takahashi KA, Jiang W, Linden DJ, Heintz N (1997) Neurodegeneration in *Lurcher* mutant mice caused by mutation in the delta 2 glutamate receptor gene. Nature 388:679–673.

IV. DEVELOPMENT AND AGING

12. DEVELOPMENT AND MOTOR CONTROL: FROM THE FIRST STEP ON

Guy Cheron, Anita Cebolla, Françoise Leurs, Ana Bengoetxea and Bernard Dan

Laboratory of Neurophysiology and Movement Biomechanics, ISEPK, Université Libre de Bruxelles, Avenue P. Héger, CP 168, Brussels 1050, Belgium and Laboratory of Electrophysiology, Université de Mons-Hainaut Department of Neurology, University Children's Hospital Queen Fabiola Brussels, Belgium

Abstract

For performing their very first unsupported steps, often considered as a 'milestone' event in locomotor development, toddlers must find a compromise between at least two requirements: (1) the postural stability of the erect posture integrating the direction of gravity and (2) the dynamic control of the body and limbs for forward progression these two aspects. In adults, a series of experimental studies have provided evidence for coordinative laws that lead to a reduction of kinematic degrees of freedom. When the elevation angles of the thigh, shank and foot are plotted one versus the others, they describe a regular gait loop which lies to a plane. The plane orientation and the loop shape reflect the phase relationship between the different segments and therefore the timing of intersegmental coordination. The general pattern of intersegmental coordination and the stabilization of the trunk with respect of vertical are immature at the onset of unsupported walking in toddlers, but they develop in parallel very rapidly in the first few weeks of walking experience. Adult-like cross-correlation function parameters were reached earlier for shank-foot pairs than for thigh-shank indicating disto-proximal maturation of the lower limb segments coordination. We also demonstrated that a dynamic recurrent neural network (DRNN) is able to reproduce lower limb kinematics in toddler locomotion by using multiple raw EMG data. In the context of motor learning the DRNN may be considered as a model of biological learning mechanisms underlying motor adaptation. Using this artificial learning during the very first steps we found that the attractor states reached through learning correspond to biologically interpretable solutions.

Introduction

Human motor repertoire can be diviided into two classes, gross and fine motor behaviour. The first classe involves the skifull use of the whole body including mobility and posture, whereas the second classe involves the use of individual body parts mainly head and hands in goal directed movements. In human, the upright position of the body has permit a full expression of upper limb movement extremities but has it the same time render the postural task more problematic by the restriction of the sustantation base. Standing and balance functions must work in conjunction in order to constantly assume antigravity muscle contraction to hold the body in upright position and to maintain the projection of the centre of gravity in the sustentation base avoiding falling over. With respect to this conservative postural task the displacement of the body is assumed by rhythmic or cyclic motor activity which is mainly organized by a "central pattern generator"(CPG) (Grillner and Zanger 1975) localized in the lumbosacral spine (Deliagina et al. 1983; Dimitrijevic et al. 1998; Yakovenko et al. 2002).

The CPG is also considered as a more general neural network co-ordinating the activity of multiple muscle into postural synergies (Forssberg and Hirschfeld 1994). These authors proposed a CPG with two functionnal levels: the first level selects the robust muscle activation pattern whereas the second level finely tuned the selected pattern by multisensorial input from visual, vestibular and somatosensory systems. The two levels CPG model is supported by experimental data on postural responses during sitting (Hadders-Algra et al. 1996). They demonstrated that the expression of the first CPG level occurs before the infant is able to sit independently (5–6 month-old) and provide a large repertoire of direction specific responses from which the most appropriate patterns are selected. Ontogenic evolution may occurs within this CPG level by improved trigerring action of afferent and/or supraspinal influences. The activity of the second CPG level emerges at about 9 months of age and is able to finely modulate the amplitude of the postural synergies.

The classical approach to motor development consists in a *follow up* of the emergence of motor and sensory abilities since the very first days of live. It describes the different behavioural states which are numerous and present diverse evolutions including extinction (e.g. the disappearance of the stepping reflex due to body mass increases (Thelen 1984; Schneider et al. 1990), reinforcement or bifurcation.

This *follow up* approach also comprise precise studies of different parameters of movement reflexly or voluntarily elicited. However, this approach is confronted to the redundancy of effective movements, first pointed out by Bernstein (1967). In fact, the human motor system is mechanically complex and can make use of a large number of degrees of freedom. Moreover, this classical approach is faced to the problem of "contex conditionned variability" (Tuller et al. 1982). During development motor systems show remarkable adaptability and flexibility in the presence of changing biomechanical properties of motor organs and when faced with different environmental conditions or tasks. For example, a given innervational state does not have a fixed movement consequence (e.g. the pectoralis major changes its role as a function of the angle of its pull with respect to the axis of the joint). Because of these problems it is difficult to establish the follow up of a precise motor event along a long period of time. What is the relevant event or movement parameter among the large number of movement in a full motor repertoire? How we can be sure that the studying event conserves the same nature along time and that it can be considered as the corresponding primitive of the mature event?

One way to partly avoid these problems is first to define in the adult movements some coordinative principles and to look backward in children toward their point of emergence. The present Chapter tempt to demonstrate the usefulness of this later approach and then to scrutinize the EMG-kinematics relationships by means of a dynamics recurrent neural network (DRNN).

Emergence of a Coordinative Template in Human Locomotion

Our understanding of the emerging coordinative principles in toddlers may benefit from recent advances in the study of walking kinematics. Mathematical approaches, ranging from neuromodulation of coupled oscillators (Kopell 1995), to synergetics (Thelen and Smith 1994), and topological dynamics (Das and McCollum 1988 ; McCollum et al. 1995), have described gait in either continuous or discrete space,

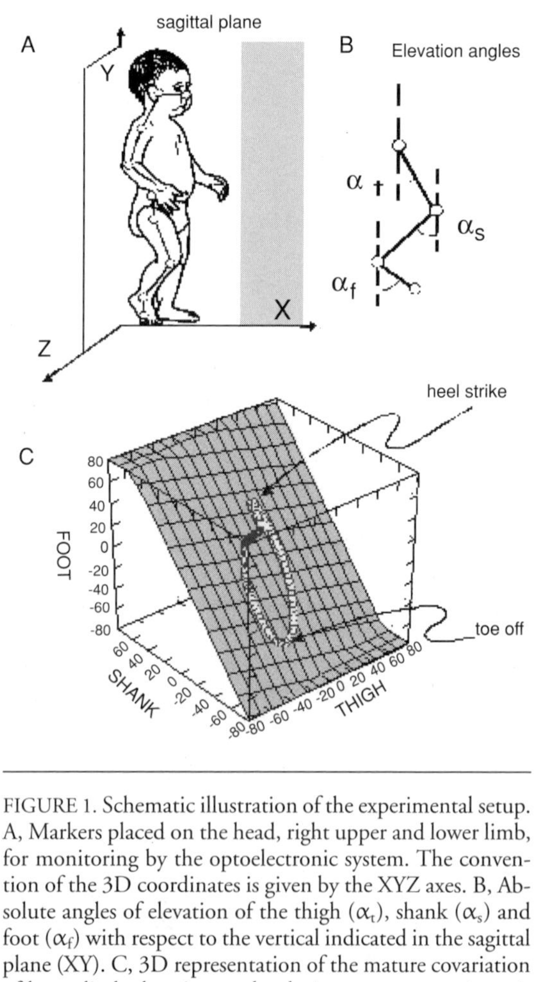

FIGURE 1. Schematic illustration of the experimental setup. A, Markers placed on the head, right upper and lower limb, for monitoring by the optoelectronic system. The convention of the 3D coordinates is given by the XYZ axes. B, Absolute angles of elevation of the thigh (α_t), shank (α_s) and foot (α_f) with respect to the vertical indicated in the sagittal plane (XY). C, 3D representation of the mature covariation of lower limb elevation angles during two consecutive gait cycles in a 12 year-old child, characterized by a quasi-elliptic loop progressing in the counter-clockwise direction and lying on a plane (grid). (*From Cheron et al 2001, Exp Brain Res. Reprinted by permission*)

and suggested that excess degrees of freedom are constrained by the neural control. As a result, limb dynamics would be confined to an attractor space of lower dimensionality than that of the original parameter space. In adults a series of experimental studies has provided detailed evidence for coordinative laws that lead to a reduction of kinematic degrees of freedom (Borghese et al. 1996; Bianchi et al. 1998a,b; Grasso et al. 1998; 1999; 2000; for a review, see Lacquaniti et al. 1999). The temporal waveform of the elevation angles of the lower limb segments (pelvis, thigh, shank and foot) relative to the vertical is much more stereotypical across trials, speeds, and subjects than the corresponding waveform of either the joint angles (Borghese et al. 1996; Grasso et al. 1998) or the EMG patterns (Grasso et al. 1998, 2000). When the elevation angles of the thigh, shank and foot are plotted one versus the others, they describe a regular gait loop which lies close to a plane (Fig. 1). The plane orientation and the shape of the loop reflect the phase relationship between the different segments and therefore the timing of intersegmental coordination (Bianchi et al. 1998b), on which postural stability with respect to gravity and dynamic equilibrium for forward progression depend. The plane orientation shifts in a predictable way with increasing speed of walking (Bianchi et al. 1998b) and with the walking posture adopted (Grasso et al. 2000). Moreover it reliably correlates with the mechanical energy expenditure (Bianchi et al. 1998a,b). The pattern of a 12-year old child is plotted in Fig. 1C. The walking cycle progresses in the counter-clockwise direction, heel strike and toe-off roughly corresponding to the top and bottom of the loop, respectively.

Developmental Emergence of the Planar Covariation in Toddlers and Children

Recently, we have characterized the developmental emergence of the planar covariation in toddlers and children (Cheron et al. 2001a,b). Figure 2 shows

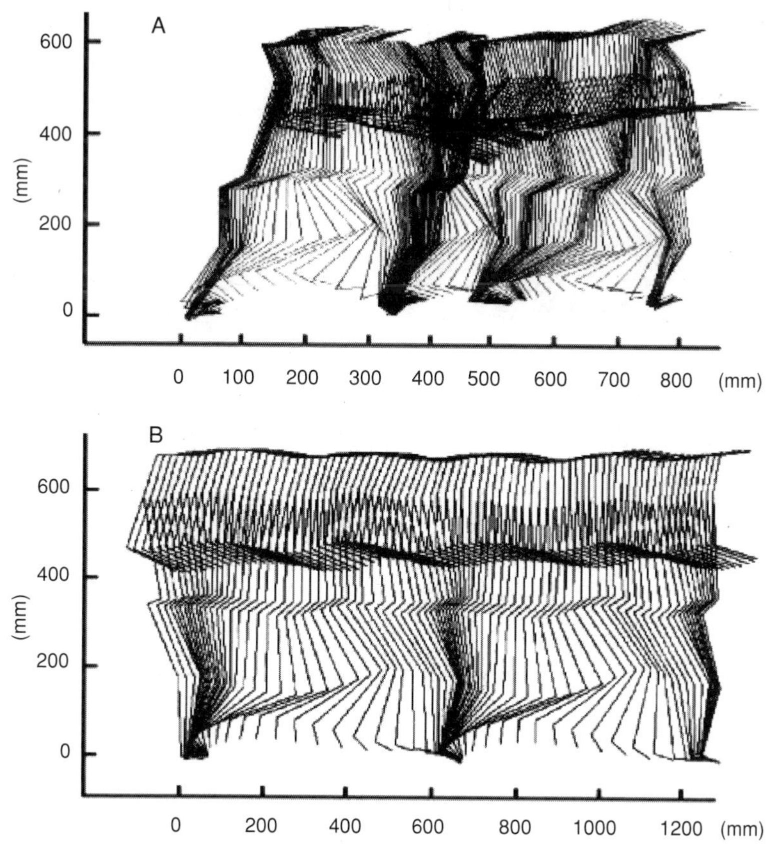

FIGURE 2. Sagittal stick diagrams at two stages of early walking. A, Very first three steps of an 11 month-old toddler. B, Two steps of the same child aged 20 months. (*From Cheron et al 2001, Exp Brain Res. Reprinted by permission*)

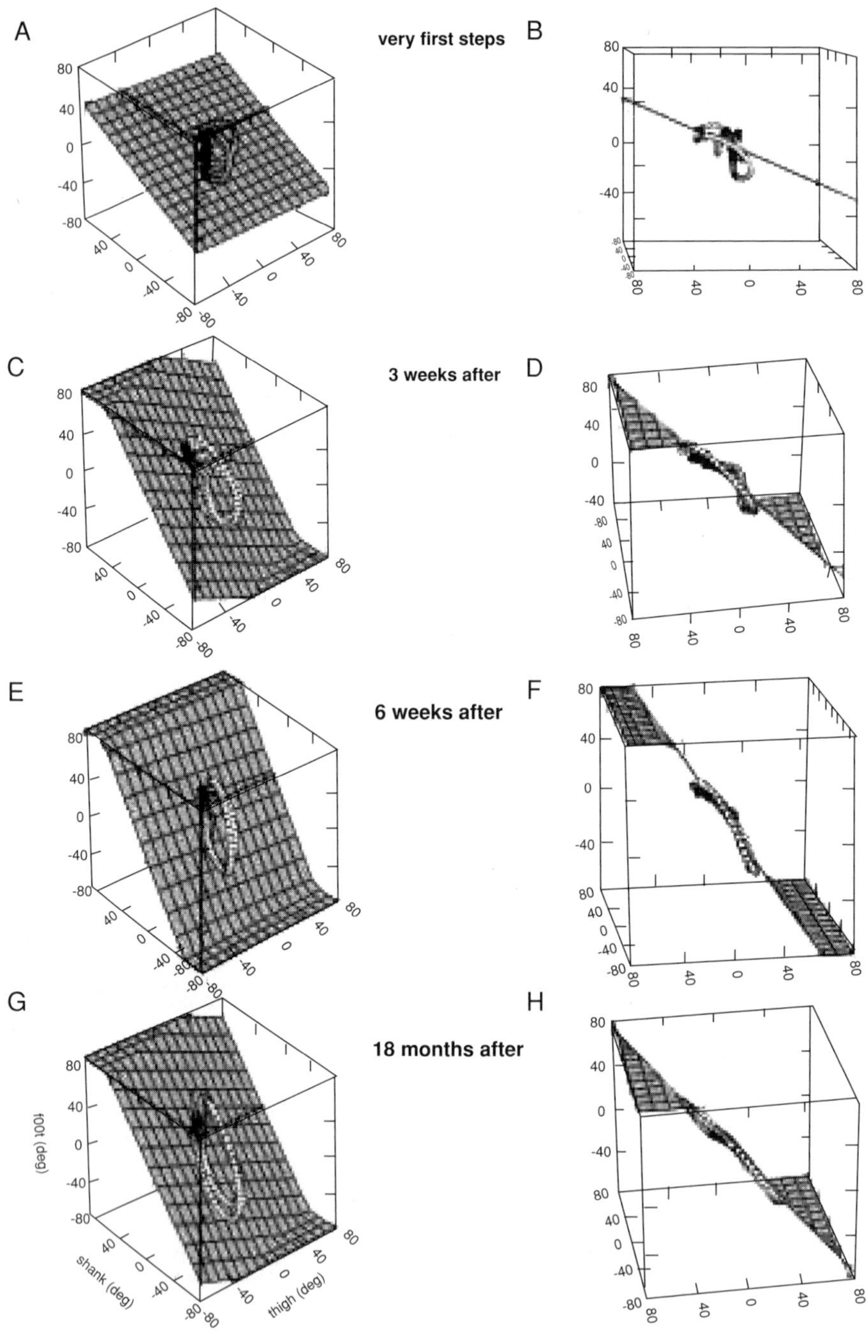

FIGURE 3. Emergence of the planar covariation of elevation angles of the thigh, shank and foot. Covariation of thigh, shank and foot elevation angles during two successive gait cycles performed by the same toddler at the onset of unsupported walking at the age of 14 months (A,B), 3 weeks later (C,D), 6 weeks later (E,F) and 18 months later at the age of 32 months (G,H). Mean value of each angular coordinate has been subtracted. The data with respect to the cubic frame of angular coordinates and the best fitting plane (grids) are represented in two different perspectives (A,C,E,G) and (B,D,F,H). Gait cycle paths progress in time in the counter-clockwise direction, heel strike and toe-off phases corresponding roughly to the top and bottom of the loops, respectively (see also Fig. 1C). (*From Cheron et al 2001, Exp Brain Res. Reprinted by permission*)

the stick diagram of the very first three steps of an 11 month-old toddler (A) and two steps performed by the same toddler at 20 months (B). The first step kinogram is characterized by a more curved trajectory of the foot associated with higher elevation of the thigh, and a larger length of the step as compared with the following steps. The trunk presents a forward sway during the initial part of the swing phase followed by a backward sway initiated well before the onset of the stance phase. This latter movement of the trunk is accompanied by neck hyperextension culminating in the middle of the swing phase. In contrast, at 20 months, during the swing phase, thigh elevation is smaller corresponding to a less marked hip flexion, hip extension occurs at the end of the stance phase and trunk sway is minimal. Head orientation in the sagittal plane is much better stabilized than at 11 month. Figure 3 illustrates the evolution of the inter-segmental coordination in one child, from her very first steps at the age of 14 months (Fig. 3A,B) to the age of 32 months (Fig. 3G, H). During the very first steps the gait loop in 3D space departs significantly from an elliptic shape (Fig. 3A), and the data are not well fitted by a plane (Fig. 3B). A planar covariation emerges early on during the following weeks of walking experience (Fig. 3C,D), and is stabilized afterwards. Note however that the shape of the gait loop matures much more slowly, with a progressive elongation of the loop along an axis roughly orthogonal to the thigh (Fig. 3E, G). This trend is related to the progressive reduction of the amplitude of thigh movement relative to that of shank and foot.

The emergence of the planar covariation rule can be discussed in relation to the neural attractor hypothesis. The idea is that the nervous system settles into preferred activation patterns, whether hard-wired or not (Koppell 1995; Thelen and Smith 1994). Such an activation pattern may depend on interaction with the physical environment, but as neural in nature it can directly control movement. The early emergence of the kinematic coordination suggests that it constitutes a specific response to dynamic functional demands imposed by human gait. The planar loop could result from a dynamical process by which a high-dimensional system compresses the many degrees of freedom involved in the realization of gait down to a low-dimensional system (Thelen and Smith 1994; Schöner et al. 1990).

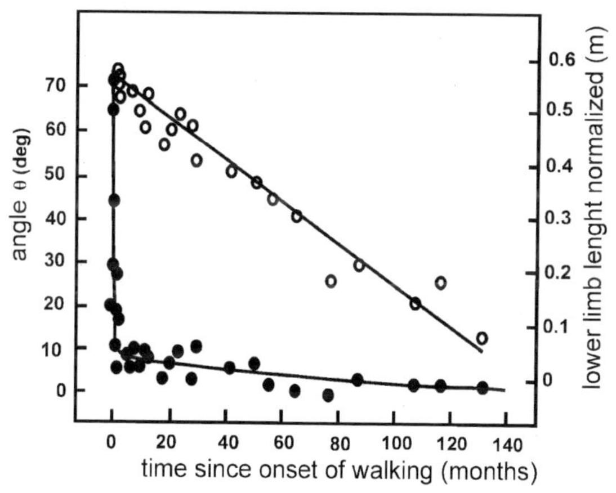

FIGURE 4. Comparison between the age changes of the covariation plane orientation and the corresponding changes of the lower limb length. For each subject and trial the angle θ between the subject's covariation plane and the mean adult plane (closed circles) and the lower limb length normalized to the mean adult value (open circles) are represented as a function of time since the onset of unsupported walking. The adult components of the plane normal are: $u_3\alpha_t = 0.223 \pm 0.092$; $u_3\alpha_s = -0.772 \pm 0.026$ and $u_3\alpha_f = 0.587 \pm 0.042$. A biexponential function and a linear regression are fitted to the angle θ values and the lower limb length respectively (see results section for more details). (*From Cheron et al 2001, Exp Brain Res. Reprinted by permission*)

Contrasting Maturation of Plane Orientation and Anthropometric Parameters

The orientation of the planar covariation represents an important parameter of the inter-segmental coordination, because it reflects the phase relationship between the different segments (Bianchi et al. 1998b). As seen in Fig. 3, the plane orientation in toddlers changes drastically over the first weeks of walking experience. These changes were quantified and compared with the changes in child morphology.

Filled points in Fig. 4A correspond to the angle (θ) between the best-fitting plane in each child and the mean adult plane. The overall time course of changes with age can be described by a biexponential function ($y = a^{-x/t_1} + b^{-x/t_2}$), where x is the time since onset of unsupported walking, t_1 is the fast time constant and t_2 is the slow one. The function fits reasonably well the experimental data ($r = 0.89$). The first time constant is fast ($t_1 = 0.59$ months after the onset of unsupported locomotion) and the orientation of the plane rapidly converges toward the adult values.

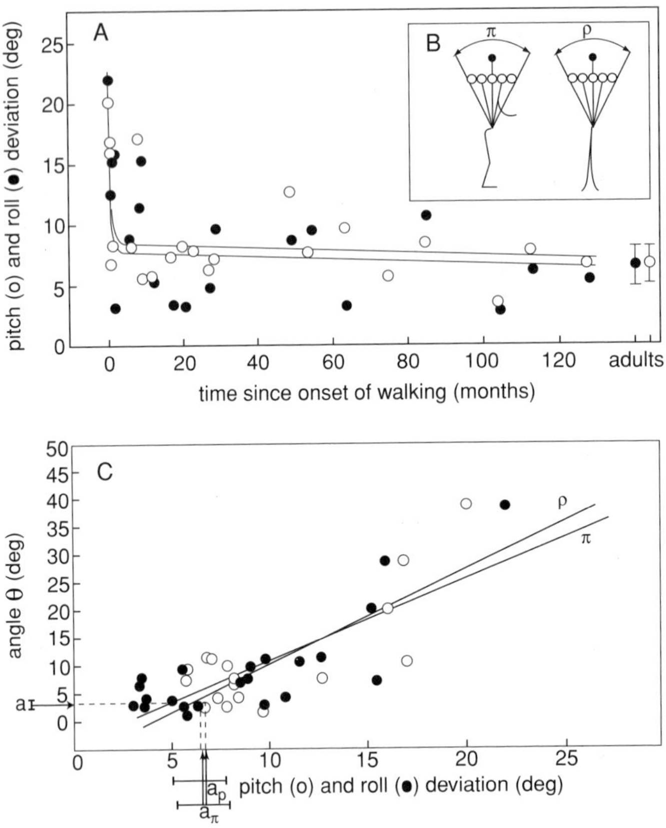

FIGURE 5. Evolution of trunk stability. A, Evolution of pitch (π) and roll (ρ) oscillations of the trunk. Age is from the onset of unsupported walking. B, Schematic definition of pitch (π) and roll (ρ) peak to peak oscillation. C, Relationship between angle θ and pitch (π) and roll (ρ) angles, with correlation coefficients (r) of 0.86 and 0.80 for θ-ρ and θ-π relationships, respectively. Adult means (stripped line) and standard deviations (I) are indexed for angle θ (a), and pitch (a_πp and roll (a_ρ) angles. (*From Cheron et al 2001, Exp.Brain Res. Reprinted by permission*)

We considered the age-related changes of two anthropometric parameters: the length of the lower limb (thigh plus shank length) normalized by the adult mean length (0.863 ± 0.055 m, unfilled points in Fig. 4), and the ratio of the lower limb length over the child stature (ear to malleolus marker distance). In contrast with the biphasic time course of changes of plane orientation, with a first quick phase, the maturation of both the lower limb length and the limb length/stature ratio is monophasic and slow.

Developmental Correlation between Trunk Stability and Planar Covariation

Analysis of trunk oscillations showed rapid stabilization in both frontal (ρ) and sagittal (π) planes (Fig. 5B). Initial peak to peak ρ and π oscillations were relatively high (14.0 ± 7.2 deg and 13.6 ± 5.8 deg, respectively, Fig. 5A). Subsequent evolution tended toward adult values (mean ρ and π = 6.4 ± 1.7 deg and 6.6 ± 1.4 deg, respectively). As for the evolution of angle θ, biexponential functions were calculated for ρ and π (r = 0.75 and 0.73, respectively) using the mean value of each angle at time 0. For both angles, the fast time constants (t_1 = 0.36 and 0.34 months after the onset of unsupported locomotion) were roughly comparable to that obtained for θ angle (0.59 months). A significant correlation (r = 0.81) was found between ρ and π trends. Figure 5C shows the existence of a significant correlation between θ and π (r = 0.80) and between θ and ρ (r = 0.86).

In healthy adults, the orientation of the covariation plane has been demonstrated to be directly related to mechanical energy cost (Bianchi et al. 1998b).

Because of the body mass distribution, trunk stability plays a determining role in mechanical energy expenditure (Bianchi et al. 1998a). It could be expected that children would approach a kinematic pattern that minimizes energy expenditure as they approach adulthood. Improvement of the covariation plane in treated patients with Parkinson's disease (Grasso et al. 1999) or hereditary spastic paraparesis (Dan et al. 2000a) also suggests a parallel improvement of the mechanical efficiency. Similarly, the correlation we found between the covariation plane orientation relative to the adult one and trunk oscillations supports the idea that the mechanical efficiency of locomotion is sustained by a highly specific orientation of the covariation plane. The bulk of the current evidence indicates that the planar covariation results from the integration of neural control and biomechanical factors. It may emerge from the coupling of neural oscillators between each other and with limb mechanical oscillators. Muscle contraction intervenes at variable times to re-excite the intrinsic oscillations of the system when energy is lost.

Maturation of stepping patterns has been shown to begin long before the child can walk (Forssberg 1985; Thelen 1985; Thelen and Cooke 1987; Yang et al. 1998) and go on long thereafter (Berger et al. 1984; Brenière and Bril 1998; Cavagna et al. 1983; Cioni et al. 1993; Clark and Phillips 1993; Forssberg 1985, 1999; Lasko-McCarthey et al. 1990; Ledebt et al. 1995; Leonard et al. 1991; Sutherland et al. 1980). This is reflected by the gradual acquisition of gait parameters, some of them as early as in fetal life (De Vries et al. 1984), some as late as late childhood (Hirschfeld and Forssberg 1992). A basic problem in maturational studies is to define the limits of a mature pattern (Forssberg 1985; Dietz 1992; Hadders-Algra et al. 1996). These limits depend on the considered parameters. For example, Bril and Brenière, (1992, 1998) have proposed two phases for walking maturation. The first phase, from 3 to 6 months after the onset of independent walking, is devoted to gait postural requirements (dynamic equilibrium during forward propulsion) and the second one, lasting about 7 years, corresponds to fine tuning of gait. Our results also support the existence of a two phase-process, as demonstrated by the biexponential evolution of the covariation plane orientation and trunk stabilization. The second phase expressed in our data by the slow time constants of the biexponential evolutions represents fine tuning, which matures more gradually than does the first phase. Other authors consider that gait maturation is finalized by the age of 7 to 8 years, through fine tuning of kinematic parameters (Sutherland et al. 1980), muscle activation patterns (Okamoto and Kumamato 1972; Berger et al. 1984; Woollacott and Jensen 1996), ground reaction forces (Gomez Pellico et al. 1995), head control and coordination (Assaiante and Amblard 1993) or anticipatory postural adjustments (Hirchfeld and Forssberg 1992; Ledebt et al. 1998). However, other gait parameters may require an even longer time to reach maturation (Cheron et al 2001a,b).

A Dynamic Recurrent Neural Network for Human Locomotion Studies

The majority of neural networks used for EMG-to-kinematics mapping have been of the feedforward type (Sepulveda et al. 1993; Koike and Kawato 1994). In these networks, information flows, without any feedback connection, from the input neurones to the output neurones. This excludes context and historical information, which are thought to be crucial in motor control (Kelso, 1995). In contrast, recurrent neural networks take these aspects into account and are recognised as universal approximators of dynamical systems (Hornik 1989; Doya 1993). Therefore, they seem particularly relevant to the study of motor control (Draye 2001; Draye et al. 2002).

Figure 6 illustrates the input-output relationships of the DRNN. The central circle represents the whole connectivity of the DRNN. Each EMG signal is sent to all the 20 artificial neurones (hidden unit) which converge to 3 output units acting merely as summation units. Each output neurone provides one specific type of kinematic data (in the illustrated situation: the angular velocity of the thigh, shank and foot).

Successful learning was ascertained on the basis of the comparison between the DRNN output and the actual output (provided by experimental data). Figure 7 illustrates the superimposition of these data (Fig. 7B,C,D) when the training has reached an error value of 0.001. The learning performance was examined on-line by inspection of the error curve (Fig. 7A). The learning process was carried out for 5000 iterations. This procedure was recently used and proved useful for the study of the very first step in toddlers. In this case the learning is also possible but with respect to the adult, the percentage of success learning decrease significantly and the number of iteration needed for reaching an error value of 0.001 increase. These difficulties for the artificial learning of the very first step may be explained by the presence of a larger amount of co-activation EMG pattern in toddlers (Fig. 8A) in comparison to the highly reciprocal activation patterns recorded in adult (Fig. 8B).

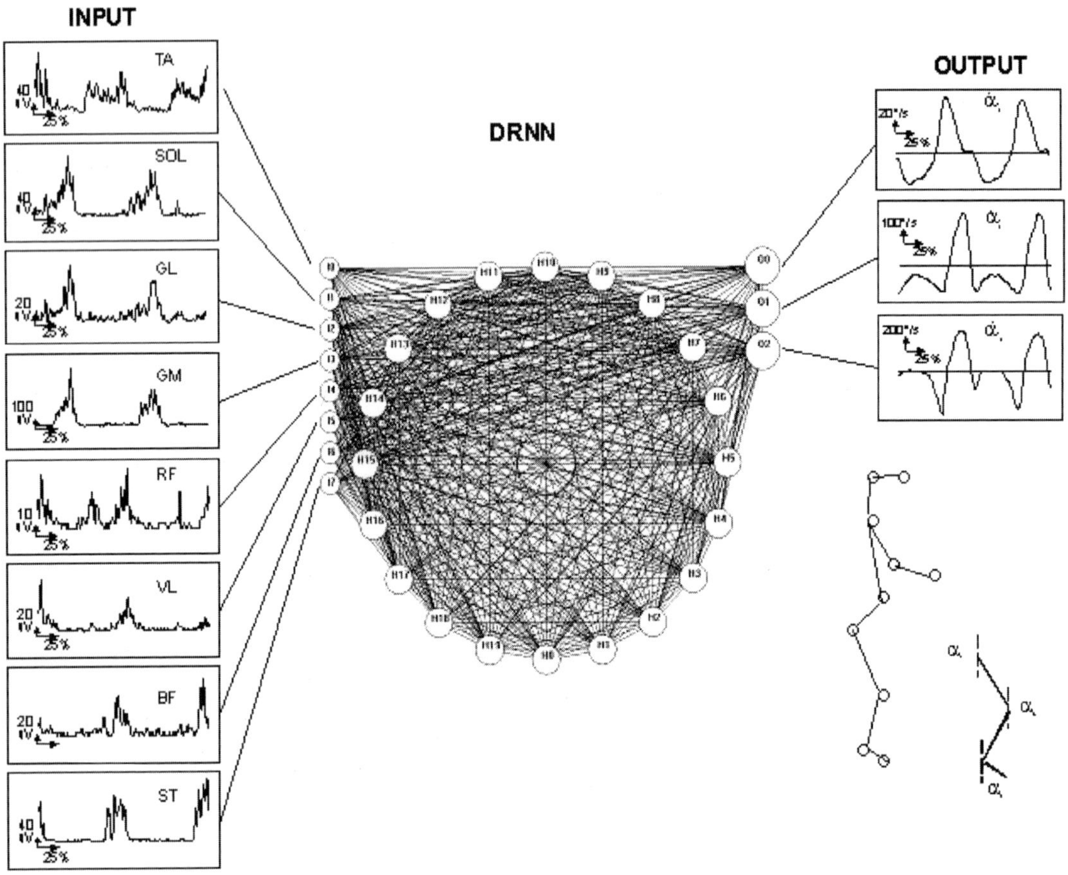

FIGURE 6. Input-output relationships of the DRNN. The central box symbolises the DRNN. Each EMG signal is sent to all 20 artificial neurones (hidden unit) which converge to 3 output units acting merely as summation units. Each output neurone provides one specific type of kinematic data represented by the absolute angles of elevation of the thigh (α_T), shank (α_S) and foot (α_F) with respect to the vertical as indicated in the stick diagram of the insert. The open circles represent the placement of the passive markers.

Biological Plausibility and Developmental Issue of DRNN Approach

In spite of the problems encountered in the EMG to kinematics mapping in toddler we have tested after the learning phase the physiological plausibility of the DRNN identification. The basic idea was to compare the angular directional change induced by artificial EMG potentiation of a single muscle with the physiological knowledge of the pulling direction of the muscle (Cheron et al. 2003). This knowledge is easily accessible for mono-articular muscles, but is less straightforward for the pluri-articular muscles. In the latter, the muscle force can be involved in a force regulation process for which the directional action is not directly defined by the pulling direction of the muscle. Moreover, dynamical coupling between the three joint segments can be implicated in the evoked movement. The implications of such complex dynamical simulations of biomechanics and muscle coordination in human walking have been recently revisited by Zajac et al. (2003). For example, Figure 9 illustrates the effect of SOL and TA artificial potentiation applied throughout the walking sequence on the sagittal lower limb kinogram over 2 steps performed by an adult. Whereas the former results in digitigrade gait (explained by the pulling action of SOL) with increased knee flexion (explained by a coupling action) more marked during the swing phase, the latter results in increased ankle dorsiflexion (walking on the heel explained by the pulling

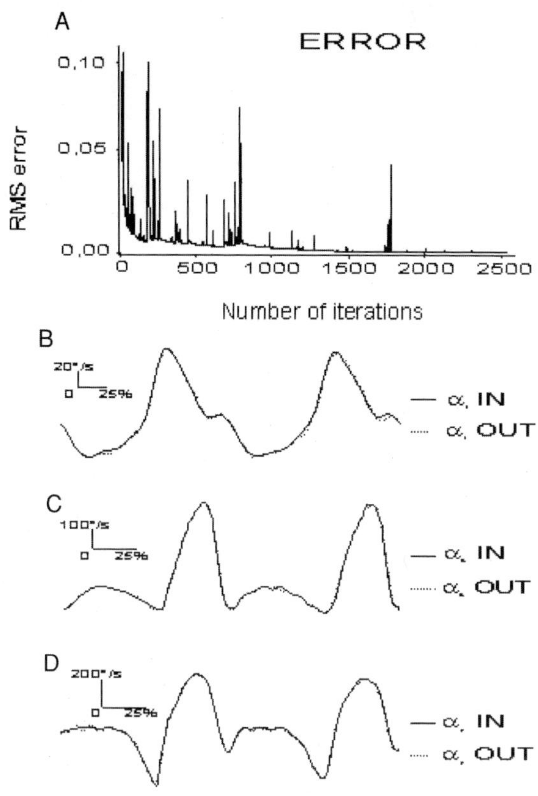

FIGURE 7. Assessment of successful learning. A, Error curve of one learning trial reaching an error value of 0.001 after 5000 iterations. B, C and D, superimposition of experimental (continuous line) and DRNN (dotted line) output signals when training reaches an error value of 0.001.

action of TA) and knee hyperextension (coupling action) more marked during the stance phase. We have also investigated the physiological plausibility of the DRNN for the very first step data by the application of a selective burst increase of the GAS (Fig. 10A) or the TA (Fig. 10B) muscle occurring during the stance phase. In both cases the resulting changes were in accordance with the physiological action of these muscles.

Toward an Integrative Tool for the Sensorimotor Coordination Dynamics

Our approach demonstrated that by using multiple raw EMG data, the DRNN is able to reproduce in adult (Cheron et al. 2003) and toddler a major parameter of lower limb kinematics in human locomotion. This dynamic mapping provides a new tool for understanding the development of the functional relationships between multiple EMG profiles and the resulting movement. In the context of motor learning the DRNN may be considered as a model of biological learning mechanisms underlying motor adaptation (Cheron et al. 1996). According to Conditt et al. (1997), adaptation to change in human movement dynamics is achieved by neuronal modules. These modules realise learning through dynamic mapping between kinematic states (positions or velocities) and the forces associated with these states. The brain is thus capable of forming and memorizing remarkably accurate internal representations of body segment dynamics (Conditt and Mussa-Ivaldi 1999). This establishes a functional relation between force and motion, which is generally complex and non-linear (Zajac and Winters 1990). Using artificial learning of the mapping between multiple EMG patterns and velocities of lower limb segments we found that the attractor states reached through learning correspond to biologically interpretable solutions. The evolution of these states could be followed during development. This neural network is also able to decipher some motor strategies using interaction torque in multijoint movements (unpublished data). For some authors, EMG patterns are a good reflection of the motor programme used by the CNS for controlling movement (Gottlieb 1993). However, for others, EMG and kinematic patterns are emergent, non-programmable properties of the system and the control signals are positional in nature (Feldman et al. 1998; Gribble et al. 1998; McIntyre and Bizzi 1993). In this controversial context the present method is not intended to propose a model for motor control based on feedforward related EMG signal for predicting kinematics. On the contrary, we propose to use the identification between EMG signals and kinematics for deciphering the complex relatonships between multiple muscular activation and the resulting movements. This dynamic identification is particularly relevant because it represents the solution for the reduction of the number of degrees of freedom and provides an idea of the controlled operation selected by the nervous system (Sporns and Edelman 1993). Moreover, they motivated behaviorally based network modelling taking into account in their architectures neurobiological principles (Draye et al. 1997, 2002) and general theory of brain function such as the theory of neuronal group selection (Edelman 1989; Reeke and Sporns 1993; Sporns et al. 2000). According to this theory, the first basic step in development is the activation of several primary neuronal groups, which are genetically determined and not definitely wired. Then, in cases of the most successful

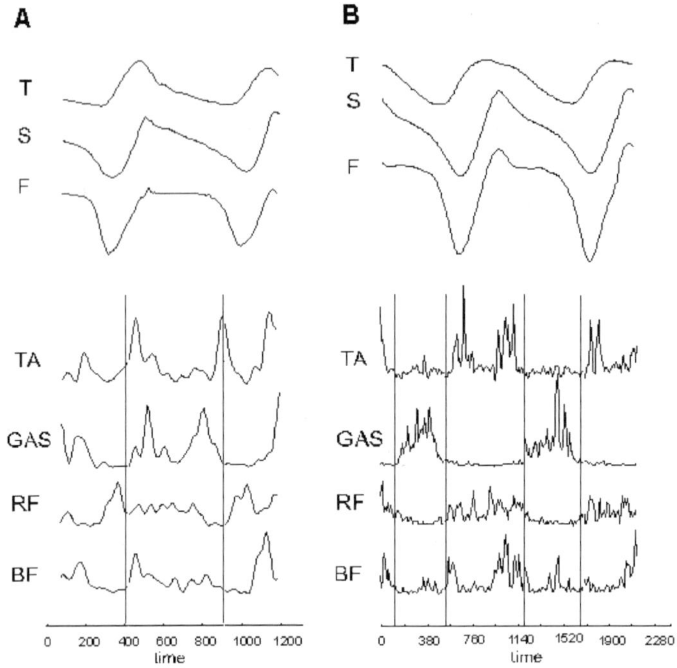

FIGURE 8. Comparison of gait activation patterns and related kinematics in a toddler (A) and an adult (B). Elevation angles of the thigh (T), shank (S) and foot (F) are illustrated in both upper parts. The related activation pattern (rectified EMG) of the tibialis anterior (TA), gastrocnemius (GAS), rectus femoris (RF) and biceps femoris (BF) are illustrated in both lower part. Note that the well characterized reciprocal EMG patterns between TA and GAS and between GAS and RF in the adult (B) are not present in the toddler (A).

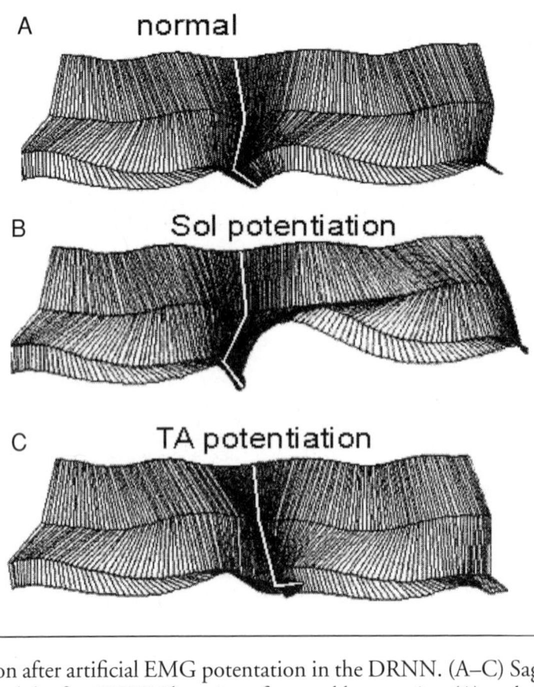

FIGURE 9. Kinematics simulation after artificial EMG potentation in the DRNN. (A–C) Sagittal stick diagrams of the lower limb kinematics obtained in an adult after DRNN learning of normal locomotion (A) and after artificial EMG potentiation of SOL (B) and TA (C) muscles. (*Reprinted by permission from Cheron et al 2003, J Neurosc. Meth.)*

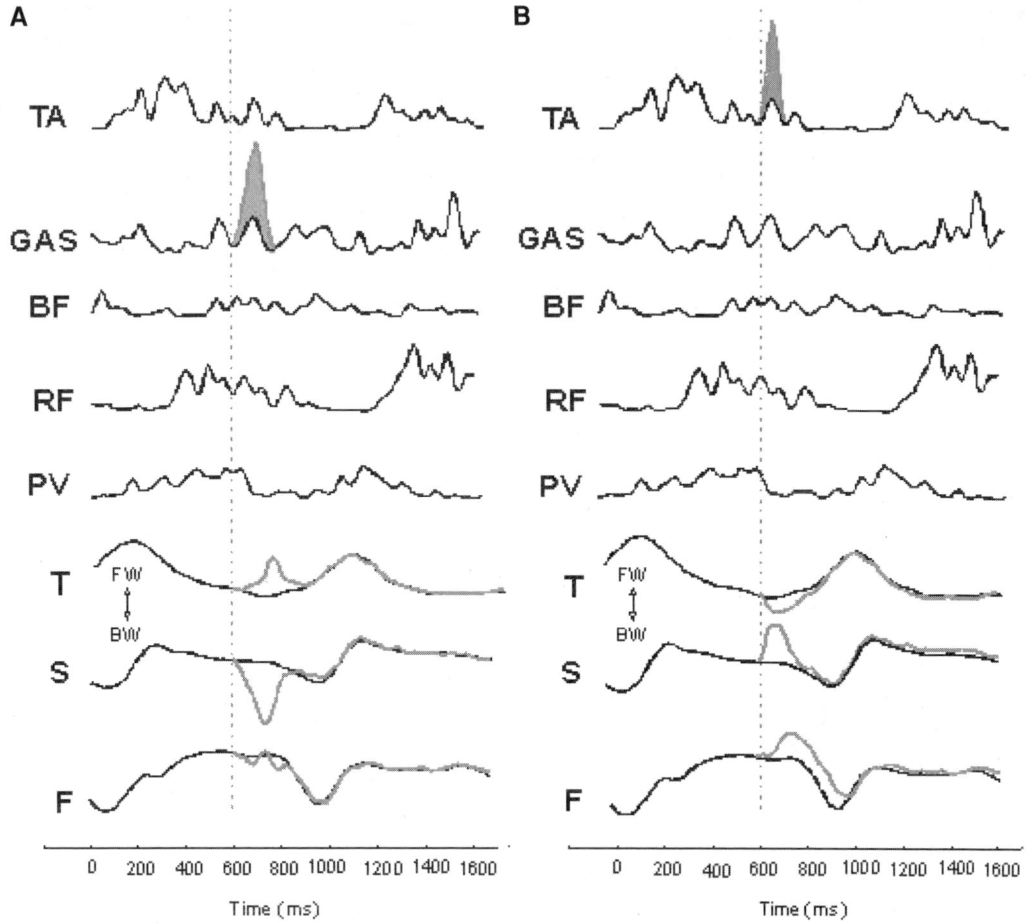

FIGURE 10. Biological plausibility of DRNN for the very first steps of a toddler. Multiple EMG patterns recorded during the very first step of a toddler and used for the DRNN learning (A,B). Artificial potentiation of GAS burst (A) and TA burst (B) resulting in ankle extension (A) (similar as those illustrated for an adult in Fig. 9B) and ankle flexion (B) (similar as those illustrated for an adult in Fig. 9C), respectively.

motor outputs, experientially driven selection occurs on these primary neural groups by synaptic reinforcement. The simulation obtained by the DRNN approach makes possible insights into how coordinated behavior is controlled by neuronal activity accessible in human on the basis of EMG recordings.

References

Assaiante C, Amblard B (1993) Ontogenesis of the head stabilization in space during locomotion in children : influence of visual cues. Exp Brain Res 93(3):499–515.

Berger W, Altenmüller E, Dietz V (1984) Normal and impaired development of children's gait. Human Neurobiology 3:163–170.

Bernstein N (1967) The coordination and regulation of movements. London: Pergamon.

Bianchi L, Angelini D, Lacquaniti F (1998a) Individual characteristics of human walking mechanics. Pflügers Arch 436:343–356.

Bianchi L, Angelini D, Orani GP, Lacquaniti F (1998b) Kinematic co-ordination in human gait: relation to mechanical energy cost. J Neurophysiol 79:2155–2170.

Borghese NA, Bianchi L, Lacquaniti F (1996) Kinematic determinants of human locomotion. J Physiol 494:863–879.

Brenière Y, Bril B (1998) Development of postural control of gravity forces in children during the first 5 years of walking. Exp Brain Res 121:255–262.

Bril B, and Brenière Y, (1992) Postural requirements and progression velocity in young walkers. J Mot Behav 24:105–116.

Cavagna GA, Franzetti P, Fuchimoto T (1983) The mechanics of walking in children. J Physiol 343:323–339.

Cheron G, Leurs F, Bengoetxea A, Draye JP, Destree M, Dan B. (2003) A dynamic recurrent neural network for multiple muscles electromyographic mapping to elevation angles of the lower limb in human locomotion. J Neurosci Methods 129:95–104.

Cheron G, Bouillot E, Dan B, Bengoetxea A, Draye JP, Lacquaniti F (2001a) Development of a kinematic coordination pattern in toddler locomotion: planar covariation. Exp Brain Res 137:455–66.

Cheron G, Bengoetxea A, Bouillot E, Lacquaniti F, Dan B (2001b) Early emergence of temporal co-ordination of lower limb segments elevation angles in human locomotion. Neurosci Lett 308:123–127.

Cheron G, Draye JP, Bourgeois M, Libert G (1996) A Dynamic neural network identification of electromyography and arm trajectory relationship during complex movements. IEEE Trans Biomed Eng 43:552–558.

Cioni G, Duchini F, Milianti B, Paolicelli PB, Sicola E, Boldrini A, Ferrari A (1993) Differences and variations in the patterns of early independent walking. Early Hum Dev 35:193–205.

Clark JE, Phillips SJ (1993) A longitudinal study of intralimb coordination in the first year of independent walking: a dynamical systems analysis. Child Dev 64:1143–1157.

Conditt MA, Gandolfo F, Mussa-Ivaldi FA (1997) The motor system does not learn the dynamics of the arm by rote memorization of past experience. J Neurophysiol 78:554–560.

Conditt MA, Mussa-Ivaldi FA (1999) Central representation of time during motor learning. Proc Natl Acad Sci U S A 96:11625–11630.

Dan B, Bouillot E, Bengoetxea A, Cheron G (2000a) Effect of intrathecal baclofen on gait control in human hereditary spastic paraparesis. Neurosci Lett 280:175–178.

Dan B, Bouillot E, Bengoetxea A, Noël P, Kahn A, Cheron G (2000b) Head stability during whole body movements in spastic diplegia. Brain Dev 22:99–101.

Das P, McCollum G (1988) Invariant structure in locomotion. Neuroscience 25:1023–1034.

De Vries JIP, Visser GHA, Prechtl HFR (1984) Foetal mobility in the first half of pregnancy. In: Prechtl HFR (eds) Continuity of Neural Functions from Prenatal to Postnatal Life. Clinics in Developmental Medicine, Spastics International Medical Publications, Oxford, pp 46–64.

Doya K (1993) Universality of fully connected recurrent neural networks. Technical report University of California: San Diego.

Draye JP (2001) Recurrent neural networks: properties and models. In: Masterbroek HAK, Vos JED, (eds) Plausible neural networks for biological modelling. Kluwer Academic Publishers: Dordrecht, 49–74.

Draye JP, Cheron G, Pavisic D, Libert G (1997) Improved identification of the human shoulder kinematics with muscle biological filters. In: Keravnou E, Garbay C, Baud R, Wyatt J (eds) Lecture Notes in Artificial Intelligence. Springer 417–429.

Draye JP, Winters JM, Cheron G (2002) Self-selected modular recurrent neural networks with postural and inertial subnetworks applied to complex movements. Biol Cybern 87:27–39.

Edelman GM (1989) Neural Darwinism. The Theory of Neuronal Group Selection. Oxford University Press, Oxford.

Feldman AG, Ostry DJ, Levin MF, Gribble PL, Mitnitski AB (1998) Recent tests of the equilibrium-point hypothesis (lambda model). Motor Control 2:189–205.

Forssberg H (1985) Ontogeny of human locomotor control. I. Infant stepping, supported locomotion and transition to independent locomotion. Exp Brain Res 57:480–493.

Forssberg H (1999) Neural control of human motor development. Curr Op Neurobiol 9:676–682.

Gomez Pellico L, Rodriguez Torres R, Dankloff Mora C (1995) Changes in walking pattern between five and six years of age. Develop Med Child Neurol 37:800–806.

Gottlieb GL (1993) A Computational model of the simplest motor program. J Mot Behav 25:153–161.

Grasso R, Bianchi L, Lacquaniti F (1998) Motor patterns for human gait: backward versus forward locomotion. J Neurophysiol 80:1868–1885.

Grasso R, Peppe A, Stratta F, Angelini D, Zago M, Stanzione P, Lacquaniti F (1999) Basal ganglia and gait control : apomorphine administration and internal pallidum stimulation in Parkinson's disease. Exp Brain Res 126:139–148.

Grasso R, Zago M, Lacquaniti F (2000) Interactions between posture and locomotion: motor patterns in humans walking with bent posture versus erect posture. J Neurophysiol 83:288–300.

Gribble PL, Ostry DJ, Sanguineti V, Laboissiere R (1998) Are complex control signals required for human arm movement? J. Neurophysiol 79:1409–1424.

Hadders-Algra M, Brogren E, Forssberg H (1996a) Ontogeny of postural adjustments during sitting in infancy : variation, selection and modulation. J Physiol 493:273–288.

Hadders-Algra M, Brogren E, Forssberg H (1996b) Training affects the development of postural adjustments in sitting infants. J Physiol 493:289–298.

Hirschfeld H, Forssberg H (1992) Development of anticipatory postural adjustments during locomotion in children. J Neurophysiol 68:542–550.

Hornik K, Stinchcombe M, White H (1989) Multilayer feedforward networks are universal approximators. Neural Networks 2:359–366.

Kelso JA (1995) Dynamic Patterns. The Self-Organization of Brain and Behavior. MIT Press Cambridge 334.

Koike Y, Kawato M (1994) Estimation of arm posture in 3D-Space from surface EMG signals using a neural network model. IEICE Trans Inf & Syst E77-D:368–375.

Kopell N (1995) Chains of coupled oscillators. In: Arbib M. A (ed) The Handbook of Brain Theory and Neural Networks. MIT Press Cambridge Mass pp, 178–183.

Lacquaniti F, Grasso R, Zago M (1999) Motor patterns in walking. News in Physiol Sci 14:168–174.

Lasko-McCarthey P, Beuter A, Biden E (1990) Kinematic variability and relationships characterizing the development of walking. Dev Psychol 23:809–837.

Ledebt A, Bril B, Brenière Y (1998) The build-up of anticipatory behaviour. An analysis of the development of gait initiation in children. Exp Brain Res 120:9–17.

Ledebt A, Bril B, Wiener-Vacher S (1995) Trunk and head stabilization during the first months of independent walking. Neuroreport 6:1737–1740.

Leonard CT, Hirschfeld H, Forssberg H (1991) The development of independent walking in children with cerebral palsy. Dev Med Child Neurol 33:567–577.

McCollum G, Holroyd C, Castelfranco AM (1995) Forms of early walking. J Theor Biol 176:373–390.

McIntyre J, Bizzi E (1993) Servo Hypotheses for the Biological Control of Movement. J Mot Behav 25: 193–202.

Okamoto T, Kumamato M (1972) Electromyographic study of the learning process of walking in infants. Electromyography 12:149–158.

Schneider K, Zernicke RF, Ulrich BD, Jensen JL, Thelen E (1990) Control of lower leg intersegmental dynamics during spontaneus kicking in 3-months-old infants. J Mot Behav 22:493–520.

Schöner G, Jiang WY, Kelso JAS (1990) A synergetic theory of quadrupedal gait and gait transitions. J Theor Biol 142:359–391.

Sepulveda F, Wells DM, Vaughan CL (1993) A neural network representation of electromyography and joint dynamics in human gait. J Biomech 26:101–109.

Sporns O, Edelman GM (1993) Solving Bernstein's problem: a proposal for the development of coordinated movement by selection. Child Develop 64:960–981.

Sutherland DH, Olshen R, Cooper L, Woo SL (1980) The development of mature gait. J Bone and Joint Surg 62:336–353.

Thelen E (1984) Developmental origins in motor coordination: Leg movements in human infants. Develop Psychobiol 18:1–22.

Thelen E, Cooke DW (1987) Relationship between newborn stepping and later walking: a new interpretation. Develop Med Child Neurol 29:380–393.

Thelen E, Smith LB (1993) A Dynamic System Approach to the development of Cognition and Action. MASS: MIT Press Cambridge.

Tuller B, Turvey MT, Fitch HL (1982) The Bernstein perspective: II. The concept of muscle linkage or coordinative structure. In: J.A.S. Kelso (ed) Human motor behavior: An introduction. Hillsdale, NJ: Erlbaum, pp 253–270.

Woollacott MH, Jensen JL (1996) Posture and locomotion. In: Handbook of motor skills. Heuer H, Keele SW (eds) Academic London, pp 333–403.

Yang JF, Stephens MJ, Vishram R (1998a) Transient disturbances to one limb produce co-ordinated, bilateral responses during infant stepping. J Neurophysiol 79:2329–2337.

Yang JF, Stephens MJ, Vishram R (1998b) Infant stepping: a method to study the sensory control of human walking. J Physiol 507:927–937.

Zajac FE, Neptune RR, Kautz SA (2003) Biomechanics and muscle coordination of human walking. Part II: Lessons from dynamical simulations and clinical implications. Gait Posture 17:1–17.

Zajac FE, Winters JM (1990) Modeling musculoskeletal movement systems: joint and body segmental dynamics, musculoskeletal actuation, and neuromuscular control. In: Winters JM, Woo SLY (eds) Multiple Muscle Systems: Biomechanics and Movement Organization. Springer-Verlag: New-York, pp 121–148.

13. CHANGES IN FINGER COORDINATION AND HAND FUNCTION WITH ADVANCED AGE

Mark L. Latash*, Jae Kun Shim*, Minoru Shinohara#, and Vladimir M. Zatsiorsky*

*Department of Kinesiology, The Pennsylvania State University, University Park, PA 16802, USA

#Department of Integrative Physiology, University of Colorado, Boulder, CO 80309, USA

Summary

Age-related changes in the hand neuromuscular apparatus are accompanied by changes in both finger strength and finger coordination. The loss of strength is more pronounced during maximal torque production tasks than in maximal force production tasks. Intrinsic hand muscles show a disproportionate loss of force, which may render multi-digit synergies learnt over the lifetime suboptimal. Age leads to lower force production by uninstructed fingers (lower enslaving), which may have negative effects on performance in tasks that involve rotational equilibrium constraints. Elderly persons show worse stabilization of the total force during accurate force production tasks (the stabilization is achieved by co-variation of forces produced by individual digits). They also show worse stabilization of the total moment produced on a hand-held object as compared to young persons. Some of the age-related changes, such as higher safety margins and higher antagonist moments produced by finger forces, may be viewed as adaptive. Other changes, however, are likely to interfere with the everyday hand function making it suboptimal.

1. *Finger Coordination as a Problem of Motor Redundancy*

The system for the production of voluntary movements is characterized by motor redundancy. This means that at any level of description the system has more elements contributing to performance than absolutely necessary to solve a motor task. Serial kinematics chains with more than three joints are redundant in kinematics while parallel kinematic chains are redundant in statics. For example, many patterns of individual joint rotations of the arm can produce a certain trajectory of the endpoint of the limb (Mussa-Ivaldi et al. 1988) while in multi-finger grasps many combinations of the finger forces can produce the desired net force and moment on a hand-held object (Li et al. 1998). Similarly, a value of joint torque does not define a unique combination of activation levels of muscles crossing the joint and many patterns of motor unit recruitment can produce a certain level of activation of a given muscle (cf. Latash 1996). The controller, the central nervous system (CNS) seems to be always confronted with a problem of choice: How to select a particular way of solving each particular problem? From a mathematical standpoint, problems of this type are ill-posed; in the motor control area they have been commonly addressed as the Bernstein problems (Turvey 1990). Bernstein himself viewed the problem of "elimination of redundant degrees-of-freedom" as the central issue of motor control (Bernstein 1947, 1967).

The hand is arguably the most versatile human effector. It is also a very attractive object to address the problem of motor redundancy. Hand function requires cooperation of the five digits towards motor goals. At the interface with a hand-held object, the digits produce forces and moments of forces that sum up to generate a required motor effect. Typically, as we will demonstrate later, the number of mechanical variables that describe the action of the digits of the human hand is higher than the number of variables specifying a task. Hence, an infinite number of combinations of digit forces and moments can satisfy virtually any task. This is a typical problem of motor redundancy. During force production and prehensile tasks, forces and moments produced by individual digits can be recorded with high accuracy thus making multi-digit action an attractive object to address the Bernstein problem.

There have been two major approaches to the Bernstein problem. The first follows the traditions set by the mentioned Bernstein's formulation and searches for a unique solution for each problem of motor redundancy. This is commonly done by adding constraints to the system or selecting a cost function and optimizing its value (reviewed in Latash 1993; Rosenbaum et al. 1995; Prilutsky, Zatsiorsky 2002). The second approach follows the traditions of Gelfand and Tsetlin (1967). It assumes that the CNS does not eliminate the degrees-of-freedom and does not select a unique solution but rather it uses all the available degrees-of-freedom to facilitate families of solutions that are equally successful to solve the task. This approach has been recently developed in the form of the uncontrolled manifold (UCM) hypothesis (Scholz and Schöner 1999; reviewed in Latash et al. 2002a). We will discuss applications of the UCM hypothesis to finger interaction studies later in this Chapter in section 3.

2. *Indices of Finger Interaction during Pressing Tasks*

During static flexion force production tasks, individual finger forces show phenomena of mutual dependence. These phenomena are interpreted as reflecting both the specific peripheral design of the hand and the neural organization of finger force control (Leijnse et al. 1993; Kilbreath and Gandevia, 1994; Roullier 1996; Latash et al. 2002b). In particular, extrinsic hand muscles such as flexor digitorum profundis (FDP) and flexor digitorum superficialis (FDS) have multiple tendons that insert at different fingers. There are also passive inter-finger links provided by connective tissue. On the other hand, finger representations in the primary motor cortex show mosaic pictures with many overlaps (Schieber 2001), an arrangement that is rather far from the perfect Penfield's homunculus (Penfield and Rassmussen 1950).

Finger interaction during force production tasks has been described using three major indices, sharing, enslaving, and force deficit (Z-M Li et al. 1998; Zatsiorsky et al. 1998). Sharing (S) reflects the fact that individual fingers typically produce certain stable percentages of the total force over a wide range of total force magnitudes. Enslaving (E) addresses unintended force production by fingers of a hand when a subset of fingers is required to produce force (Kilbreath and Gandevia 1994; Schieber 2001). Force deficit (FD) reflects the fact that a finger produces lower peak forces during multi-finger MVC tasks as compared to its peak force when it is required to produce MVC alone (Kinoshita et al. 1995; Ohtsuki 1981). Quantitatively these indices have been characterized as:

$$S_i = 100\%\ F_{i,task}/F_{tot,task}$$

$$E_{i,j} = 100\%\ F_{i,j}/F_{i,i}$$

$$FD_{i,task} = 100\%\ (F_{i,i} - F_{i,task})/F_{i,i},$$

where subscripts i and j refer to fingers (index, I, middle, M, ring R, and little, L), subscript tot stands for total, and task indicates a multi-finger task. Certain regularities have been observed in these indices across the healthy, young subjects. In particular, typically, the index and middle fingers produce about 60% of the total force, while the little finger produces only about 15%. Enslaving effects are stronger between couples of adjacent fingers and are nearly symmetrical, i.e. the magnitudes of $E_{i,j}$ and $E_{j,i}$ are close to each other. Force deficit increases with the number of fingers explicitly involved in the task.

Both extrinsic and intrinsic hand muscles are activated during many daily activities, such as grip and pinch (Darling et al. 1994). The different anatomical points of attachment of extrinsic and intrinsic muscles (Basmajian and DeLuca 1985) present an opportunity to vary the relative involvement of these muscle groups by changing the site of external force application (Danion et al. 2000; Z-M Li et al. 2000). Extrinsic flexors (FDP and FDS) are multi-digit muscles and focal flexors at the distal interphalangeal (IP) joint and at the proximal IP joint respectively, while intrinsic muscles act as digit-specific focal flexors at the metacarpophalangeal (MCP) joints in addition to their extensor action at more distal joints (Landsmeer and Long 1965; Long 1965). Hence, when a person presses with fingertips, extrinsic flexors are focal force generators while intrinsic muscles participate in balancing moments at the MCP joints. When a person presses with proximal phalanges while keeping the distal phalanges in an intermediate posture, intrinsic digit-specific muscles become focal force generators while extrinsic flexors balance the action of the extensor mechanism at IP joints (An et al. 1985; Chao et al. 1976). In particular, MVC produced at the fingertips requires peak force production by extrinsic flexors while intrinsic muscle involvement has been assessed as ranging between 10% and 30% of their MVC (Harding et al. 1993; Z-M Li et al. 2000). In contrast, when a person presses maximally by proximal phalanges, intrinsic muscles are expected to produce forces close to their MVC, while existing assessments of forces produced by the extrinsic muscles suggest that they require the two major extrinsic flexors to produce below 20% of their maximal forces (Chao and An 1978; Harding et al. 1993; Landsmeer and Lang 1965; Smith 1974).

Studies of the indices of finger interaction during force production at the two sites—distal and proximal—revealed qualitatively similar patterns of S, E, and FD at the two sites, while the magnitude of E and FD was significantly higher when the subjects produced forces at the proximal phalanges (Latash et al. 2002b). This observation has suggested that the patterns of finger interaction are mostly defined by central neural factors and do not depend crucially on the presence of multi-digit extrinsic muscles. On the other hand, a study of the effects of transcranial magnetic stimulation on finger force responses at the distal phalanges showed that the magnitude of a response in a finger depended strongly on the background force produced by this finger but showed no or weak dependence on the background force produced by other fingers of the hand (Danion et al. 2003a). These observations suggest a high degree of physiological independence among the compartments of the extrinsic flexor muscles (cf. Jeneson et al. 1992; Fleckenstein et al. 1992; Serlin and Schieber 1993; Bickerton et al. 1997).

Some of the early studies of finger interaction during pressing tasks suggested that when one finger changed its force, other fingers could also change their forces in such a way that the total force was relatively stabilized. In particular, the variance of the total peak force computed over a set of ramp force production trials was shown to be significantly smaller than the sum of the variances of individual finger forces (Z-M Li et al. 1998). This observation suggests negative covariation of individual finger peak forces. In another study, subjects were asked to produce a submaximal constant force with three fingers pressing in parallel and then to tap with one finger (Latash et al. 1998). During tapping, the finger lost contact and stopped producing force. Other fingers showed an out-of-phase change in their forces partly compensating the effects of the tapping finger on the total force. A similar finding has been reported for experiments in which subjects removed or added a finger to a set of fingers generating force (S. Li et al. 2003).

3. *Force and Moment Stabilization in Multi-Finger Tasks*

The uncontrolled manifold hypothesis assumes that, when the CNS stabilizes a particular value of a performance variable produced by an apparently redundant multi-element system, it selects a subspace within the state space of the elements such that the desired value of the performance variable is constant. This subspace has been termed the "uncontrolled manifold" (UCM). After selecting a UCM, the controller selectively restricts the variability of elements along "essential" directions within the state space that do not belong to the UCM, while directions within the UCM can show relatively high variability of the elements' outputs. In other words, the controller allows the elements to show high variability (have more freedom) as long as it does not affect a desired value of an important performance variable (hence, the term "uncontrolled manifold"). This hypothesis views motor systems as abundant rather than redundant, i.e. it views additional degrees-of-freedom not as a computational burden but as a luxury that allows motor patterns to be adaptable and flexible.

The UCM-hypothesis allows to introduce an operational definition for a multi-effector synergy. A synergy can be defined as a task-specific organization of the effectors that stabilizes a certain value or a time profile of an important performance variable. When a potentially important performance variable is selected, UCM can be computed, and the total variance (V_{TOT}) in the state space of the effectors (elements) can be decomposed into two orthogonal components, quantified per degree-of-freedom, parallel to the UCM (V_{UCM}) and orthogonal to the UCM (V_{ORT}). If the former is significantly larger than the latter ($V_{UCM} > V_{ORT}$), one may claim that the effectors' outputs co-vary to stabilize the performance variable, i.e. that there is a synergy with respect to that performance variable.

Figure 1 illustrates the notion of the UCM for a task of producing a certain value of total force (e.g., 20 N) by quickly pressing with the two index fingers on separate force sensors. Panel A shows two possible distributions of the data points (illustrated by ellipses). The spherical distribution corresponds to a non-synergy according to the introduced definition. The elliptical distribution corresponds to a synergy stabilizing the total force since the amount of variance parallel to the UCM (shown by the dashed line and corresponding

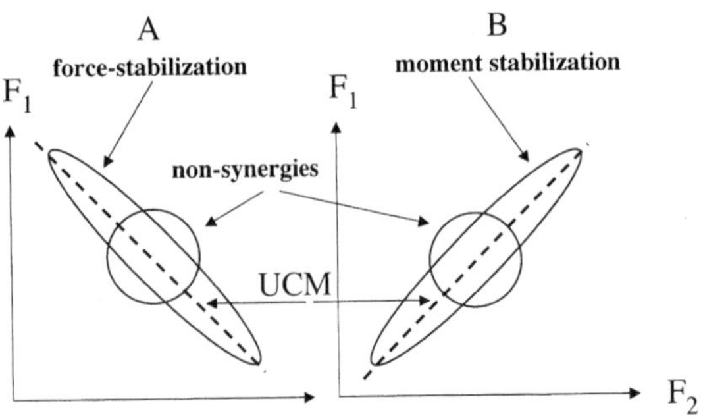

FIGURE 1. An illustration of the UCM approach using an experimnt with two-finger force production. A: The spherical distribution of data points corresponds to a non-synergy. The elliptical distribution corresponds to a synergy stabilizing the total force at 20 N. B: This elliptical distribution of data points destabilizes the total force but it stabilizes the total moment produced by the forces with respect to the midpoint between the fingers. Dashed staright lines show UCMs.

to an equation $F_1 + F_2 = 20$ N) is larger than the variance orthogonal to the UCM. Individual finger forces show predominantly negative co-variation across trials, which stabilizes the total force. Panel B shows that such a task may be associated with another UCM (dashed line) corresponding to another synergy. The positive co-variation of the finger forces destabilizes the total force but it stabilizes another important variable, the total moment produced by the finger forces with respect to the midpoint between the fingers. Note that the UCM analysis deals with relative magnitudes of variance within and orthogonal to a UCM such that a system may show high or low absolute accuracy with or without using an adequate synergy.

This type of analysis was applied to test whether a set of fingers within a hand form a synergy that selectively stabilizes a total force profile or a total moment of forces with respect to the longitudinal axis of the hand/forearm (pronation/supination moment). Note that the fingers of a hand are not independent force generators because of the mentioned phenomenon of enslaving. To overcome this problem, UCM analysis was performed using a set of hypothetical independent elemental variables termed force *modes* (Latash et al. 2001; Scholz et al. 2002; Danion et al. 2003). Force modes were defined based on control trials when the subjects were asked to produce ramp force profiles with one finger at a time. In different studies, subjects were required to produce fast and slow, ramp and oscillatory changes in the total force while pressing with 2, 3, or all 4 fingers of the hand (Latash et al. 2002c). The following major results were obtained:

1. Total force was stabilized by predominantly negative co-variation of force modes only at relatively high values of and slow changes in the total force;
2. Total pronation/supination moment was stabilized over most tasks, particularly at high rates of force production;
3. Initiation of a force ramp production was always associated with positive co-variation of force modes leading to destabilization of the total force; and
4. There was a subject-specific critical time after the beginning of a trial when fingers started to show negative co-variation of force modes and thus stabilize the total force profile (Shim et al. 2003b).

4. *Changes in the Motor Function with Age*

Aging leads to changes in many aspects of voluntary movements. These include, in particular, the slowness in movement initiation and in movement execution (Stelmach et al. 1988, 1987; Welford 1984). The internal structure of voluntary movements is changed showing longer deceleration phases (Cooke et al. 1989; Darling et al. 1989; Pratt et al. 1994), increased incidence of corrective adjustments during fast targeted movements (Pratt et al. 1994) and higher reliance on visual feedback control (Seidler-Dobrin and Stelmach 1998).

Elderly are known to be concerned about accuracy (Welford 1984) and show increased safety margins in a variety of motor tasks (e.g., Cole 1991). Time pressure is another important potential factor that may interfere with natural performance of motor tasks by the elderly (Stelmach et al. 1988, 1987; Welford 1984). If there is no time pressure, elderly use proprioceptive and sensory information similar to young persons (Chaput and Proteau 1996a). Time pressure makes elderly rely on proprioceptive information more (Chaput and Proteau 1996b).

Excessive muscle coactivation is commonly seen in elderly. In particular, during fast voluntary movements, elderly persons show scaling of electromyographic (EMG) patterns in the agonist-antagonist muscle pairs similar to that seen in young persons but with a relatively larger coactivation of the muscles (Seidler-Dobrin and Stelmach 1998). There is also a marked co-contraction of agonist-antagonist muscle groups in response to postural perturbations (Woollacott et al. 1988).

The number of alpha-motoneurons declines with age (Campbell et al. 1973). This loss becomes apparent after the age of 60. High threshold motor unit atrophy is particularly pronounced (Owings and Grabiner 1998). The relation between motor unit size and fatigability tends to break down and larger motor units (MUs) become as fatigable as smaller ones; normally large, fatigable MUs are reduced in size (see Luff 1998 for a review). Muscles lose both cross-sectional area and fiber numbers; most affected are type II fibers. This leads to a higher percentage of type I fibers (reviewed in Kirkendall and Garrett 1998; Bemben 1998). Elderly persons demonstrate increased apparent muscle stiffness (McDonagh et al. 1984), and reduced tendon compliance (Tuite et al. 1997).

Another age-related factor is loss of muscle force and mass associated with a loss of both voluntary and electrically-evoked strength (Winegard et al. 1997). Strength becomes a limiting factor in certain everyday activities such as rising from a chair (Hughes et al. 1996). There are controversial reports regarding possible differential losses of force in different muscle groups with age. In particular, some authors reported significant differences between the force loss in the upper and lower extremity muscles (Grimby et al. 1982) and between proximal and distal muscles (Nakao et al. 1989; Shinohara et al. 2003b). Other studies, however, failed to confirm these results (e.g., Viitasalo et al., 1985).

5. *Changes in Finger/Hand Control with Age*

Aging leads to a decline in hand strength and loss of manual dexterity, which affects many of the activities of daily living (Boatright et al. 1997; Giampaoli et al. 1999; Hughes et al. 1997; Rantanen et al. 1999; Francis and Spirduso 2000). This is associated with changes in the neuromuscular apparatus such as a drop in the number of motor units, an increase in the size of the motor units and a general slowing down of their contractile properties (Doherty and Brown 1997; Duchateau and Hainaut 1990; Kamen et al. 1995; Kernell et al. 1983; Owings and Grabiner 1998). Many clinical scales of motor abilities rely heavily on hand function (e.g., Jebsen Hand Function Test, Hackel et al. 1992). However, relatively few studies have addressed age-related changes in finger coordination during force and moment production tasks (Contreras-Vidal et al. 1998; Cole et al. 1999; Cole and Rotella 2002; Shinohara et al. 2003a,b).

Distal arm muscles show particularly pronounced changes with age. Thumb abduction strength, pinch strength, and grip strength all decrease after the age of 60 (Boatright et al. 1997). The index finger shows reduced abduction strength and increased force fluctuations (Galganski et al. 1993). Motor units within the first dorsal interosseus muscle show more variable discharge rates while the maximal discharge rate is reduced (Kamen et al. 1995). Studies of the first dorsal interosseus muscle have shown excessive coactivation of the second palmar interosseus and coactivation of an antagonist (Spiegel et al. 1996). One may expect that neural control of fingers adjusts to these changes to optimize hand performance in everyday motor tasks.

The decline in the overall performance of the hand within a broad range of functions is accompanied by a drop in the tactile and vibration sensitivities (Kenshalo 1979). These two processes may be related to each other: Denny Brown (1966) has reported that cutaneous sensitivity of fingertips plays a crucial role during precise manipulation. Kinoshita and Francis (1996) compared force control during prehension in young and elderly subjects. They found that elderly subjects showed lower skin friction, higher safety margins, more fluctuations in the grip force curve, and longer times of force application. Higher safety margins were also reported by Cole (1991) that could be related to changes in skin friction and/or to production of comparably strong sensory signals in elderly. In more recent studies, however, Cole and his colleagues (Cole et al. 1998, 1999) have challenged a hypothesis that the decline in the ability of older persons to grip and lift objects is solely due to their impaired tactile sensitivity. Contreras-Vidal et al. (1998) studied the performance of elderly subjects in handwriting tasks and have suggested that the spatial coordination of fingers and wrist movements declines with age while control of force pulses may be preserved. All these observations suggest that the deterioration of performance in tasks involving hand and fingers in elderly can get contribution from both peripheral and central neural factors.

A recent series of studies of the effects of aging on the structure of force variability during the isometric submaximal force production have shown that age leads to both an increase in the variability and a change in the timing structure of the force signal (Vaillancourt and Newell 2003; Vaillancourt et al. 2003).

6. *Age-Related Changes in Finger Interaction in MVC Pressing Tasks*

Our recent studies of changes in indices of finger interaction during pressing tasks have led to both expected and unexpected results illustrated in Figure 2 (Shinohara et al. 2003a, b, 2004). Expectedly, elderly persons, both males and females, showed smaller peak finger forces across the tasks as compared to younger

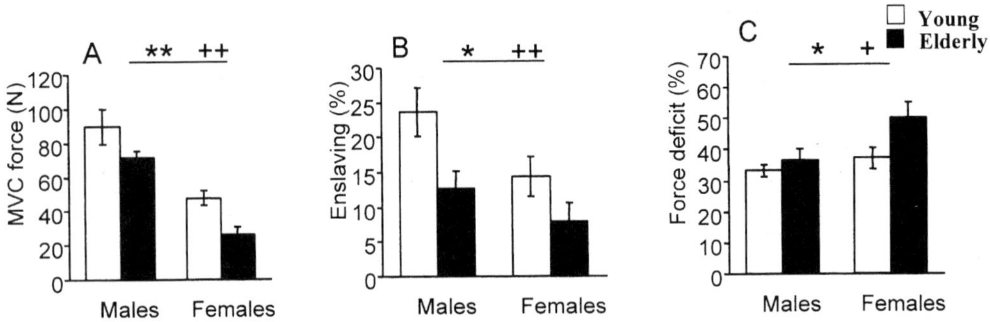

FIGURE 2. Changes in maximal force (MVC), enslaving (E), and force deficit (FD) with age. Average across subjects data are shown with standard error bars. (Reproduced with permission from Shinohara et al. 2003a).

subjects. The difference was of the order of 30% in the four-finger IMRL MVC task and it was about 20% in single-finger MVC tasks.

Surprisingly, elderly persons showed significantly lower indices of enslaving as compared to young persons. Lower enslaving can be interpreted as better individual control of finger forces or higher dexterity (cf. S. Li et al. 2000). This finding is counter-intuitive taking into account the general decline in the hand function with age. At the same time, force deficit showed higher magnitudes in elderly persons.

Connective tissue has been shown to replace contractile proteins with aging (Zimmerman et al. 1993). This could be expected to lead to an increase in enslaving due to increased 'parallel' force transmission among structures serving individual digits, not to the mentioned findings of lower enslaving in elderly. The enlargement of motor units associated with aging (reviewed in Larsson and Ansved 1995) could also be expected to lead to increased enslaving due to increased chances of simultaneous recruitment of fibers from compartments of extrinsic hand muscles serving individual digits. Hence, the finding of decreased enslaving strongly suggests changes at the level of central commands to motoneuronal pools in elderly.

Increased force deficit in elderly subjects could be due to changed motor unit properties as well as to modified supraspinal control. Force produced by a muscle is a consequence of both the number of recruited motor units and their discharge rate. Similarly, force deficit may be viewed as a consequence of both incomplete recruitment of motor units and their reduced discharge rate. Due to the increased innervation ratio of motor units with aging (e.g., Larsson and Ansved 1995 for review), a lack of recruitment of a fixed number of motor units may be expected to result in a relatively larger drop in force in elderly subjects. In addition, possible effects of reduced discharge rate of motor units on force deficit may be related to changes in the force-frequency dependence (Cooper and Eccles 1930; Thomas et al. 1991; Shinohara et al. 2003a). One can conclude, therefore, that changes in force deficit with age also suggest changes at neural levels involved in the generation of commands to hand muscles.

Changes in indices of finger interaction with age were qualitatively (and in some cases, also quantitatively) similar to those observed between male and female subjects (Shinohara et al. 2003a) and between young subjects prior to and after fatigue (Danion et al. 2000, 2001). There seems to be only one factor that changes in a similar way across the three comparisons, elderly vs. young, female vs. male, and fatigued vs. non-fatigued. This factor is the total force producing abilities. An analysis of the indices of finger interactions as functions of the total MVC force (MVC_F) confirmed that E expressed in percent of peak force increased with MVC_F while FD decreased with an increase in MVC_F. These relations are illustrated in Figure 3 for the data averaged across four groups of subjects, young males, young females, elderly males, and elderly females. The same graph also shows data points from an earlier study of the effects of fatigue on finger interaction (Danion et al. 2000).

In another study (Shinohara et al. 2003b), the relative contribution of intrinsic and extrinsic hand muscles to finger pressing force was manipulated by varying the site of force production along the finger. As mentioned earlier, the different sites of tendon attachment make intrinsic and extrinsic hand muscles involved to different degrees in tasks when flexion MVC is produced at the proximal phalanges and at the distal phalanges.

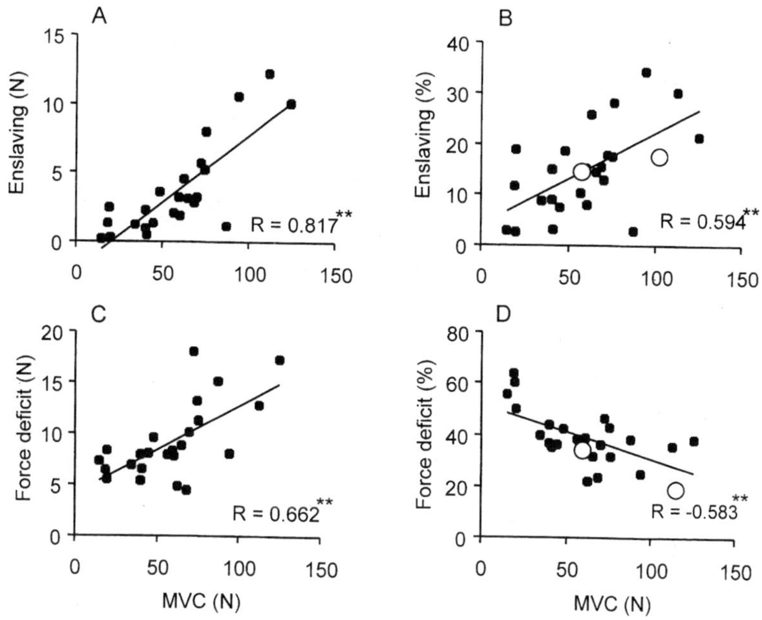

FIGURE 3. Enslaving (ENSL) and force deficit (FD) across all twenty-four subjects, male and female, young and elderly, in newtons (A) and in percent of the MVC force in its single-finger task (B) as functions of the peak force in the four-finger MVC task (MVC). Linear regression lines are shown and correlation coefficients are presented. The Figure also shows data points from an earlier fatigue study (open symbols). (Reproduced with permission from Shinohara et al. 2003a).

The decline in the peak force with age during MVC tasks was greater when the subjects performed the tasks at the proximal phalanges (30%) than at the distal phalanges (19%). These results have been interpreted as indicating a larger decline in the force producing capabilities of the intrinsic hand muscles as compared to extrinsic hand muscles. This conclusion is also supported by observations of a relatively large decline with age of the MVC force during index finger abduction task; this task requires high force production by the first dorsal interosseus, an intrinsic hand muscle (Semmler et al. 2000; Laidlaw et al. 2002; Shinohara et al. 2003b). These observations are also in a good correspondence with earlier reports on distal muscles being more affected by age than proximal muscles (Christ et al. 1992; Era et al. 1992; Viitasalo et al. 1985).

When subjects produced MVC force at the proximal phalanges they consistently showed larger indices of both enslaving and force deficit as compared with the tests at the distal phalanges. This was true across ages and genders. This observation supports the central (neural) origin of these indices of finger interaction: If the presence of multi-digit muscles were an important factor, the indices would be expected to be smaller during force production at the proximal phalanges because in those tests the focal force generators were digit-specific, intrinsic muscles.

The finding of disproportional losses of force at the two sites, proximal and distal, suggests potentially detrimental effects on muscle synergies involved in finger force production. Most everyday tasks involve force application by the fingertips. These forces generate moments in all finger joints that need to be balanced by muscle action. In particular, intrinsic muscles are required to balance moments in the metacarpophalangeal joints. Hence, commands to extrinsic and intrinsic muscles need to be accurately balanced to prevent joint motion during static tasks with fingertip force production. Such combinations of commands are probably elaborated and refined by the CNS over the lifetime based on the individual person's anatomy and the range of everyday tasks. If the force-generating capabilities of muscles involved in a synergy change disproportionately, previously developed combinations of neural commands to the muscles are likely to become suboptimal. If such changes in the muscle properties are permanent, as with aging, previously elaborated muscle synergies likely need to be adjusted. This may not be a simple task for the CNS resulting in the application of inadequate muscle synergies and decreased motor performance of the hand.

7. *Age-Related Changes in Finger Interaction in Accurate Force Production Tasks*

Most everyday tasks require accurate production of submaximal forces and force moments by the digits. It is not obvious how the demonstrated age-related impairments in the MVC tests affect performance in such tasks, more relevant to the everyday activities. A series of studies addressed multi-finger coordination during accurate force production tasks in both young and elderly persons (Latash et al. 2002c; Shinohara et al. 2003a, 2004; Shim et al. 2003a, 2004).

When a person presses on a set of force sensors with the four fingers of a hand and produces an accurate profile of the total force under continuous visual feedback, finger forces show certain patterns of co-variation both along a trial and across trials. Analyses of such co-variation patterns have been performed by comparing time patterns and average indices of the total force variance ($V_{TOT}(t)$) and the sum of individual finger force variances ($\sum V_i(t)$). The difference between the two indices, $\Delta V(t) = \sum V_i(t) - V_{TOT}(t)$, reflects prevalence of either negative co-variations among the finger forces (when $\Delta V > 0$) or positive co-variations among the forces ($\Delta V < 0$). Note that negative co-variation among the finger forces may be viewed as a force-stabilizing synergy, while positive co-variation destabilizes the profile of the total force (see also Fig. 1).

Studies in young healthy subjects have shown that, even with sufficient practice, humans cannot stabilize the total force from the very beginning of a trial while they show such force stabilization over later segments of the force ramp (Latash et al. 2002c). In a study with changes in the rate of force increase, negative co-variation among finger forces emerged only after a certain, subject-specific time delay that could range from 130 ms to over 800 ms (Shim et al. 2003b). In another study with the production of very quick force pulses, the subjects showed negative force co-variation after about 50 ms from the initiation of the trial (Latash et al. 2004). Such short time delays are probably incompatible with using sensory feedback to organize a force-stabilizing synergy and are more likely to involve short-delay central back-coupling circuits.

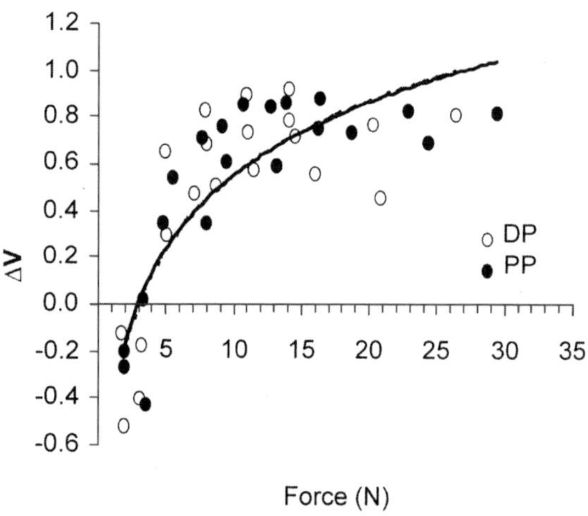

FIGURE 4. The normalized difference (ΔV) between the sum of the variances of individual finger forces and the variance of the total force during force production at the distal (open circles) and proximal phalanges (filled circles) are plotted against the actual mean force in each ramp segment for all four subject groups. The best-fit logarithmic curve is also shown. (Reproduced with permission from Shinohara et al. 2004).

The first study of the performance of elderly subjects in such tests (Shinohara et al. 2003a) showed that both young and elderly subjects showed predominantly positive co-variation among finger forces during the initial segment of the ramp. Negative finger force co-variation was seen after the total force reached a level close to 5 N (Figure 4). This common "critical force" magnitude was observed across subject groups, which differed quite dramatically in their force producing abilities. A conclusion has been drawn that this common critical force could reflect the fact that multi-finger synergies are elaborated by all persons, irrespective of their force producing capabilities, during everyday tasks that involve manipulation of objects with similar inertial properties.

Application of the UCM analysis to accurate force production tests has shown additional differences between young and elderly persons (Shinohara et al. 2004). To remind, this analysis operates with independent hypothetical variables, force modes, and it could be applied to test different hypothesis, in particular those of total force stabilization and pronation/supination moment stabilization by co-variation of force modes to individual fingers. In the analysis of force variance profiles, the magnitude of the total force when negative values of ΔV turned into

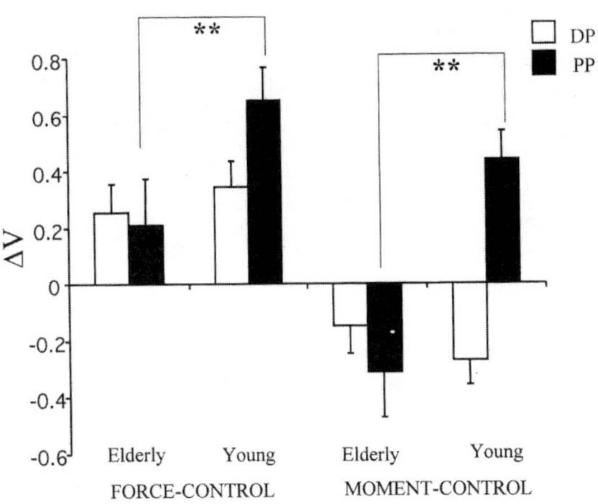

FIGURE 5. The index of covariation of finger modes (ΔV) computed for the force-control and moment-control hypotheses for the force application at distal (DP) and proximal phalanges (PP). Young subjects show higher values of ΔV as compared to elderly subjects during force application at PP but not at DP. Mean values with standard error bars are shown. (Reproduced with permission from Shinohara et al. 2004).

positive values was about the same (about 4–5 N) across the subject groups and sites of force production. Within the UCM analysis, however, additional age-related differences have been revealed. Elderly subjects took more time and reached higher forces before they were able to co-vary modes to stabilize the total force. Young subjects also showed better moment stabilization than elderly. Age-related differences in both force- and moment-stabilization effects were particularly strong during force application at the proximal phalanges when intrinsic hand muscles were the focal force generators. During force production at the proximal phalanges (Fig. 5), young subjects showed co-variation of modes that stabilized both total force and total pronation/supination moment ($\Delta V > 0$), while elderly subjects showed worse force stabilization and failed to stabilize the moment ($\Delta V \leq 0$). This observation lends additional support to the earlier conclusion on a more severe impairment of the intrinsic hand muscles with age.

This series of studies have led to a conclusion that the drop in MVC is accompanied in elderly subjects with worse coordination of control signals to fingers in multi-finger tasks (cf. Ikeda et al. 1991; Cavanaugh et al. 1999; Cole and Rotella 2002). The UCM analysis was more powerful as compared to analysis of force variance profiles in revealing significant differences between the groups.

8. *Prehensile Tasks: Mechanics and Control*

When a person grasps with five digits and manipulates a hand-held object, he or she should control simultaneously six mechanical variables per digit since each digit exerts three force components and a moment of force in the plane of the contact, while the point of application of finger force can move over the area of contact in two dimensions. We will address these as elemental variables. A stable performance with respect to an overall mechanical variable, such as the total force or the total moment of forces exerted on the hand-held object, is possible only if a spontaneous change in one of the elemental variables is compensated by coordinated changes in other elemental variable(s). A prehension synergy can be defined as a conjoint change of elemental variables during multi-finger prehension tasks (Zatsiorsky et al. 2002).

Studies of prehension synergies used external perturbations (Cole and Abbs 1987, 1988), correlations among output variables in single trials (Santello and Soechting 2000), and changes in the task parameters such as the object geometry and the resisted torque (Zatsiorsky et al. 2002). In particular, Cole and Abbs (1987, 1988) studied rapid pinch movements of the index finger and the thumb from an open-hand position and found that the finger and the thumb behaved synergistically as a single unit. Santello and Soechting (2000) reported that, within a single trial, individual normal finger forces oscillated synchronously and, hence, were determined by a common multi-finger synergy. Zatsiorsky and his colleagues (2002) showed conjoint changes in finger forces and points of their application with changes in the external force and torque.

For the system to be at rest, the sum of all forces and moments acting on the handle should equal zero. Hence, the following three requirements should be satisfied:

(1) The sum of the normal forces of the four fingers equals the normal force of the thumb

$$F^n_{th} = F^n_i + F^n_m + F^n_r + F^n_l = \sum_{f=1}^{4} F^n_f \quad (1)$$

(2) The sum of the digit tangential forces equals the weight of the hand-held object

$$L = F^t_{th} + F^t_i + F^t_m + F^t_r + F^t_l \quad (2)$$

(3) The total moment produced by the digit forces is equal and opposite to the external torque exerted on the objects

$$T = \underbrace{F^n_{th}d_{th} + F^n_i d_i + F^n_m d_m + F^n_r d_r + F^n_l d_l}_{\text{Moment of the normal forces} \equiv T^n} + \underbrace{F^t_{th}r_{th} + F^t_i r_i + F^t_m r_m + F^t_r r_r + F^t_l r_l}_{\text{Moment of the tangential forces} \equiv T^t} \quad (3)$$

where the subscripts *th, i, m, r* and *l* refer to the thumb, index, middle, ring, and little finger, respectively; the superscripts *n* and *t* stand for the normal and tangential force components, respectively; *L* is load (weight of the object), *T* is total moment or torque, and coefficients *d* and *r* stand for the moment arms of the normal and tangential force with respect to a preselected center, respectively. The equations (1)–(3) impose three constraints on the fifteen variables (normal and tangential finger force components and the coordinates of the points of force application in the vertical direction). Hence, the system has twelve degrees-of-freedom that can be manipulated by the performer in different ways. The importance of the third constraint, which unites all the elemental variables has been emphasized (Shim et al. 2003a).

It has been well established that the normal forces exerted on a hand-held object are coordinated to prevent the slipping of the object from the hand (reviewed in Johansson, 1996). Analysis of digit forces has commonly been performed within the hypothesis on the hierarchical control of prehension (Mackenzie and Iberall 1994; Iberall 1997; Baud-Bovy and Soechting 2001, 2002; Zatsiorsky et al. 2002c). According to that hypothesis, there are at least two levels of control. The first level defines the forces and moments produced by the thumb and by the virtual finger—an imaginable finger whose mechanical action is equivalent to the combined action of the actually involved fingers of the hand. The second level distributes the action of the virtual finger among the actual fingers. When the handle is oriented vertically, the normal forces of the thumb and the virtual finger have been shown to change in synchrony (Santello and Soechting 2000); they are modulated by the weight of the object (Hager-Ross et al. 1996), gravity changes during parabolic flights (Hermsdorfer et al. 1999), abrupt vertical load perturbations (Eliasson et al. 1995), tangential pulling forces (Burstedt et al. 1999), friction conditions (Edin et al. 1992; Cole and Johansson 1993), and forces acting during fast movements (Flanagan and Wing 1997; Weeks et al. 2002).

Our recent studies have shown that when people repeat a simple task of holding a handle with a certain combination of the external load and external torque, elemental variables produced by individual finger vary significantly, while their combined mechanical effect remains highly stable (Shim et al. 2003a). This is achieved by fine adjustments of forces across digits. The same study has also led to a conclusion that the control of prehension can be described by interactions within two subsets of the elemental variables. The first subset includes normal forces of the thumb and the virtual finger. The second subset includes five variables: tangential forces of the thumb and virtual finger, the moments produced by the tangential and normal forces, and the moment arm of the normal force. The compensated variations within each of the two subsets can be seen as necessitated by the task mechanics. Conjoint variations of the variables of the first subset prevent the object from slipping out of the hand and from movement in the horizontal direction. Conjoint variations among the variables of the second subset maintain the torque magnitude constant and prevent the object from moving in the vertical direction.

Although relations between the two subsets of variables are mechanically possible they are not realized. So, one can conclude that the central nervous system forms two null spaces using the two subsets of elemental variables. This finding supports the principle of superposition for human prehension that has recently been suggested for the control of prehension in robotics (Arimoto et al. 2001a,b). An entire task is divided into subtasks such that independent controllers specify different subsets of control parameters. Effects of commands from the controllers, for instance the 'torque' and 'force' commands to the digits, are added without interfering with each other. Such a control sharply decreases computation time. It is compatible with a view that the prehension synergy repesents two sub-synergies realizing correspondingly grasp control (preventing an object from slipping out of the hand) and torque control (maintaining a desired object orientation). It is worth mentioning that an overwhelming majority of the research on grasping has dealt only with the first sub-synergy (Burstedt et al. 1997; Cole et al. 1999; Flanagan et al. 1999) while the second one has typically been overlooked.

Additional support for the principle of superposition in human prhension has been obtained in experiments that varied the magnitudes of the external torque and load independently (Zatsiorsky et al. 2003). This study showed highly significant effects of both external load and external torque factors on each of the elemental variables. However, there were no significant interactions effects between these two factors suggesting the additive action of two commands related to the external load and torque.

9. *Age-Related Changes in Prehensile Tasks*

Many everyday tasks such as eating with a spoon, drinking from a glass, and writing with a pen require precise control of both forces and moments of forces produced by the digits and acting on the hand-held object. If this control is impaired, the drink will be spilled, the food will make a mess, and the pen will leave a poorly discernible scribble on the paper. To study possible age-related changes in the coordination of elemental variables produced by individual digits, we analyzed performance of subjects in static maximal and accurate submaximal force and moment production tasks (Shim et al. 2004). Elderly and young subjects pressed on six-dimensional force sensors affixed to a handle with a T-shaped attachment. The attachment allowed applying different external torques while the weight of the system was counterbalanced with another load using a pulley system.

During tasks that required the production of maximal force (MVC_F) or maximal torque (MVC_T) by all the digits, young subjects were stronger than elderly. A greater age-related deficit was seen in the MVC_T tests (Fig. 6). In particular, as compared to the young males, elderly males showed, on average, a 33.9% smaller MVC_T and only a 15.4% smaller MVC_F, a more than two-fold difference. A smaller difference was seen between the elderly and young females in the two tests: As compared to young females, elderly females showed a 23.9% smaller MVC_T and a 19.9% smaller MVC_F.

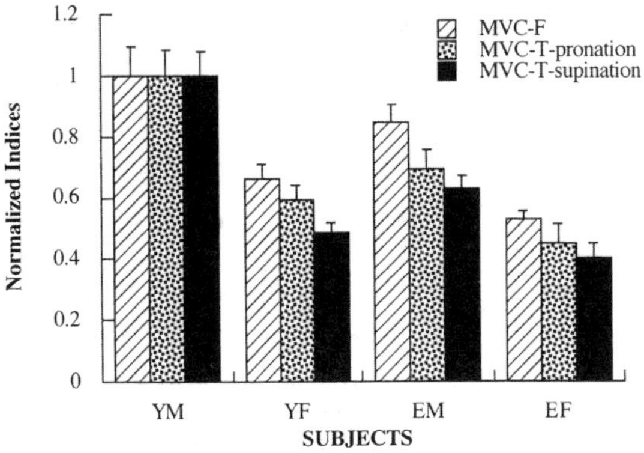

FIGURE 6. Maximal forces (MVC_F) and maximal moments (MVC_T) normalized by the mean performance of the young male subjects. YM, YF, EM, and EF represent young male, young female, elderly male and elderly female subjects, respectively. Means and standard error bars are presented. (Reproduced with permission from Shim et al. 2004.)

Several factors could have contributed to the additional decline in the performance of MVC_T tasks by elderly. First, elderly subjects produced higher forces by fingers that generated moments directed opposite to the required direction of moment production, for example index and middle finger forces for the task of moment production in supination. Such antagonist moment production in submaximal prehension tasks was reported earlier and interpreted as a consequence of enslaving, which leads to unintended force production by antagonist fingers as a result of intended commands to agonist fingers (Zatsiorsky et al. 2003). However, aging has been shown to lead to a drop in enslaving (Shinohara et al. 2003a,b) casting doubt on this interpretation.

Second, changes in the relative involvement of individual fingers could have affected the peak moment values. Elderly subjects showed a larger involvement of the index and middle fingers in the four-finger MVC_F task as compared to young subjects. This observation contrasts the earlier report on unchanged sharing patterns in the pressing tasks with age (Shinohara et al. 2003a). The difference may be due to the difference in the tasks and associated mechanical requirements: During the prehension MVC_F task in the current study, the subjects were required to maintain the orientation of the T-shaped handle system, i.e. an additional requirement of rotational equilibrium was imposed. There was also a significant difference between the elderly and control subjects in the change of the point of the thumb force application. As compared to the young subjects, elderly participants rolled the thumb up, closer to the index and middle fingers. This increased the lever arms of the forces produced by the little and ring fingers and decreased the lever arms for the other two fingers. This can be viewed as an adaptive strategy to compensate partly for the relatively lower involvement of the ring and little fingers in the elderly. The difference in MVC_T between the subject groups was indeed larger during supination tasks, when the little and ring fingers produce moments in the required direction, but it was also present in pronation tasks.

Third, the maximal moment production task (MVC_T) may be viewed as more complex and less intuitive than MVC_F. However, MVC_T was performed using the fixed handle, which did not need to be stabilized, while the MVC_F task was performed using the T-shaped handle system, which was free to move, i.e. with the additional requirement of moment stabilization. The MVC_F task was therefore associated with two mechanical requirements, maximal total force and unchanged total moment, while the MVC_T task had only one requirement, maximal total moment.

All three factors could have contributed to the observed greater impairment of the maximal moment production the elderly. Additional factors could also include the documented drop in the tactile and vibration sensitivities (Kenshalo 1979) with age. It is possible that excessive involvement of fingers that produce antagonist moments could be due to changes in skin friction and/or to production of comparably strong sensory signals in elderly (Cole 1991).

Two tasks required the accurate production of the total force and moment simultaneously, the task of holding the handle system against a non-zero external torque and zero external load (constant moment production) and the ramp force production task while keeping the orientation of the handle system constant. In both tasks, elderly subjects showed less accurate performance as quantified by the RMS error index computed with respect to both total force and total moment. These observations are in a good correspondence with earlier reports on the lower accuracy and higher variability in force production tasks by elderly subjects (Burnett et al. 2000, Enoka et al. 2003). Our findings extend these reports to moment production tasks.

To analyze possible sources of the less accurate force production by the elderly, we used an approach described earlier: We compared the sum of the variance profiles of the forces produced by individual fingers to the variance of the sum. This comparison showed prevalence of negative co-variation among the finger forces starting from the very beginning of the trial. This result contrasts the earlier reports of predominantly positive co-variations of finger forces early in the ramp trial during pressing tasks (Latash et al. 2001, 2002a,b). In another study, it has been suggested that the central nervous system needs a certain time (between 150 and 850 ms) to establish a task-specific negative co-variation of finger forces in such tasks (Shim et al. 2003). There is an important difference between the pressing and prehensile tasks: The former starts with all the fingers fully relaxed, while the latter starts with the fingers producing a non-zero background force and acting against an external torque. Our results show that, if the fingers are already involved in a synergetic activity, the CNS can organize their adequate interaction from the very beginning of the force ramp trial.

Young subjects showed higher indices of negative finger force co-variation over the whole duration of the ramp indicating better finger force coordination to stabilize the time profile of the total force (Fig. 7). In an earlier study, a similar result was observed during four-finger pressing tasks (Shinohara et al. 2004). Taken together, the studies shows that the impairment in finger coordination in elderly persons persists in prehension tasks that can be considered more relevant to everyday hand function.

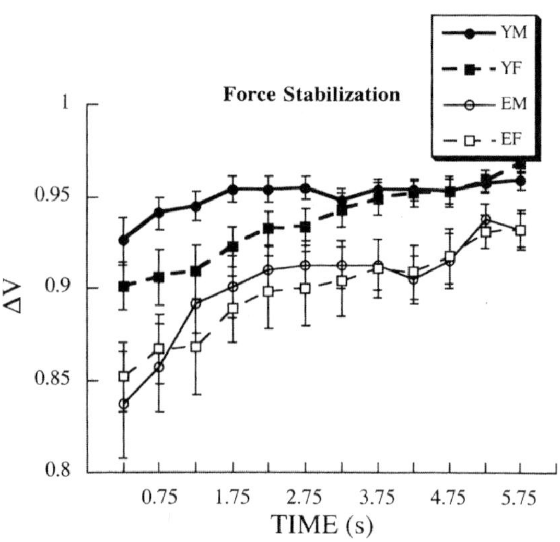

FIGURE 7. Normalized difference between the sum of the variances of the individual finger forces and the total force variance [$\Delta Var_F(t) = (\sum VarF_i(t) - VarF_{TOT}(t))/\sum VarF_i(t)$] computed over 12 trials during the ramp force production. Averages over 0.25 s time intervals are shown with standard error bars. YM, YF, EM, and EF represent young male, young female, elderly male and elderly female subjects, respectively. (Reproduced with permission from Shim et al. 2004.)

A similar analysis was run to assess the co-variation of the two components of the total moment produced by the tangential and normal digit forces respectively, M^t and M^n. This analysis has also shown higher indices of negative co-variation between M^t and M^n in young subjects as compared to elderly subjects throughout the ramp trial (Fig. 8). Hence, we can conclude that elderly subjects are impaired in their ability to organize both co-variation of forces produced by individual digits and co-variation of moment components in a task specific way.

10. *Adaptive Motor Control in Elderly*

In one of the earlier studies (Shinohara et al. 2003a), we suggested an adaptation hypothesis which implies

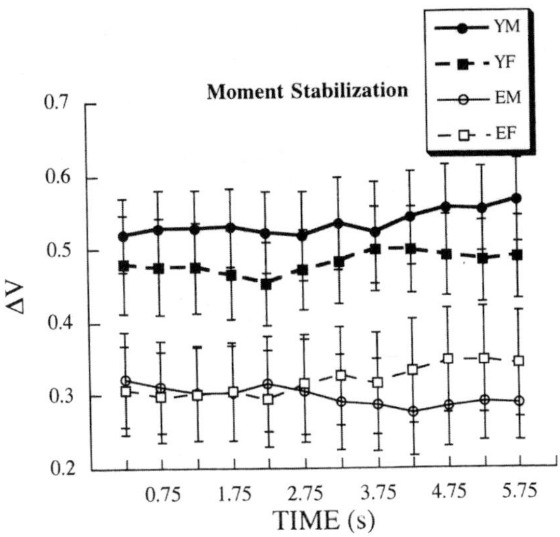

FIGURE 8. Normalized difference between the sum of the variances of the moments produced by the normal and by the tangential forces and the variance of the total moment produced by the digits $\Delta Var_M(t) = [(\sum VarM^{n,t}(t) - VarM_{TOT}(t)) / \sum VarM^{n,t}(t)]$ computed over 12 trials during the ramp force production. Averages over 0.25 s time intervals are shown with standard error bars. YM, YF, EM, and EF represent young male, young female, elderly male and elderly female subjects, respectively. (Reproduced with permission from Shim et al. 2004.)

that a loss of the muscle force, whether due to aging or fatigue (Contreras-Vidal et al. 1998), leads to changes in neural control whose purpose is to optimize the functioning of the hand across functionally important everyday tasks. Note that changes in the muscle properties with fatigue and with age show similarities including slowing of the contractile properties, which could lead to an increase in the slope of the force-frequency relation (Binder-MacLeod and McDermond 1992; Kamen et al. 1995). The steep portion of the force-frequency curve is steeper after fatigue in flexor pollicis longus (Era et al. 1992) and in quadriceps femoris (Bigland-Ritchie et al. 1986). Besides, a reduction in the maximal discharge rate of motor units have been observed under muscle fatigue (Bigland-Ritchie et al. 1983), resembling changes that occur with age (Harding et al. 1993; Miller et al. 1993).

Many of the observations reviewed in this Chapter support the adaptation hypothesis. In particular, a drop in the enslaving may be viewed as contributing to better individual control of fingers, although not without a price, since higher enslaving may be helpful in prehension tasks that involve stabilization of an object grasped by the hand (Zatsiorsky et al. 2002a,b).

The well established increase in the safety margin used by elderly persons in grasping tasks (Cole 1991; Kinoshita and Francis 1996; Gilles and Wing 2003) may also be viewed as adaptive. Aging is typically associated with increased tremor and higher variability of movement patterns (Galganski et al. 1993; Enoka et al. 2003). Both these factors contribute to poorly controlled inertial forces that may be acting on a hand-held object. Applying higher grip forces seems like a sensible strategy to assure that, even if an unexpected inertial force emerges, the increased safety margin will prevent the object from slipping out of the hand.

In our experiments, elderly subjects also demonstrated excessive grip forces, even in conditions when the grip force was not necessary because the load was zero (the weight of the handle system was counterbalanced by the counter-load). In these conditions, the non-zero grip force could partly result from the other task component, the production of a non-zero moment and from the enslaving effects (cf. Zatsiorsky et al. 2002a). Excessive grip forces by the elderly could be related to their higher moments produced by antagonist fingers, i.e. by fingers that produced moment opposite to the required moment direction. The production of excessive antagonist moments implies stronger central commands sent to those fingers. Since the total moment was to be equal to the external torque, commands to all four fingers were likely increased resulting in the higher grip force.

Both higher grip forces and higher antagonist moments may be viewed as energetically suboptimal but leading to more stable performance. Higher grip forces would prevent the object from slipping out of the hand if the load force changes, for example, due to acceleration of the object in the vertical direction or due to the variability of the grip force. Both could be expected from the less steady performance by the elderly (Burnett et al. 2000; Cole 1991; Enoka et al. 2003). On the other hand, antagonist moments can be viewed as increasing the apparent stiffness of the hand, i.e. its passive resistance to small variations in the applied torque. Overall, the results indicate that elderly subjects use higher safety margins with respect to possible variations in both force and torque. Such patterns may be viewed not as abnormal but as adaptive to the overall decline in the control of finger forces and moments. Recent studies have suggested that age-related changes in the neuromotor apparatus are accompanied by adaptive changes in the control strategies that help alleviate the detrimental effects (DeVita and Hortobagyi 2000; Shinohara et al. 2003a).

However, the finding of disproportional losses of force at the two sites, PP and DP, suggests potentially detrimental effects on muscle synergies involved in finger force production. Most everyday tasks involve force application by the fingertips. These forces generate moments in finger joints that need to be balanced by muscle action. In particular, intrinsic muscles are required to balance moments in the MCP joints. Hence, commands to extrinsic and intrinsic muscles need to be accurately balanced to prevent joint motion under fingertip force production. Such combinations of commands are probably elaborated and refined by the central nervous system (CNS) based on the individual person's anatomy and the range of everyday tasks.

If the force-generating capabilities of muscles involved in a synergy change disproportionately, previously elaborated combinations of neural commands to the muscles are likely to become suboptimal. If such changes in the muscle properties are permanent, as with aging, previously elaborated muscle synergies likely need to be adjusted. This may not be a simple task for the CNS resulting in the application of inadequate muscle synergies and decreased motor performance of the hand (Connelly et al. 1999; Grabiner and Enoka 1995; Shinohara et al. 2003a).

One may suggest two ways of dealing with this problem. First, massive practice may help the CNS revise the inadequate muscle synergies and elaborate new ones. However, the continuing changes in the muscle properties with age may prevent the CNS from elaborating new optimal sets of commands to hand muscles. Alternatively, efforts can be directed at restoring the balance between the force-generating capabilities of the intrinsic and extrinsic muscles. This goal may be more realistic with the help of specifically focused training programs.

Effects of training have been documented in many studies of elderly subjects. In particular, training has been shown to lead to higher forces and lower antagonist coactivation. Since muscle cross-sectional area showed only minor enlargements in the process of training, neural adaptations were likely to play a major role in bringing about these effects (e.g., Hakkinen et al. 1998). A recent report has suggested that the impaired ability of elderly to control pinch force accurately can be improved with specialized training (Ranganathan et al. 2001). It remains to be seen whether tasks that require coordination of digits to produce combinations of force and moment can also show improvement with practice in elderly. This is a challenge for future studies.

Acknowledgments

This study was supported in part by NIH grants AG-018751, NS-035032, AR-048563, and M01 RR10732. We are grateful to Ning Kang, Brendan Lay, and Sheng Li for their help in performing the experiments at different stages of this project, to the personnel of the General Clinical Research Center at The Pennsylvania State University for screening the subjects, and to the staff and participants at the Foxdale Village (State College, PA) for their cooperation.

References

An KN, Kwak, BM, Chao EY, Morrey BF (1984) Determination of muscle and joint forces: a new technique to solve the indeterminate problem. J Biomech Eng, 106: 364–367.

Arimoto S, Nguyen PTA (2001) Principle of superposition for realising dexterous pinching motions of a pair of robot fingers with soft-tips. IEICE Trans Fundament Elec Comm Comp Sci E84A: 39–47.

Arimoto S, Tahara K, Yamaguchi M, Nguyen PTA, Han HY (2001) Principles of superposition for controlling pinch motions by means of robot fingers with soft tips. Robotica 19: 21–28.

Atkeson CG (1989) Learning arm kinematics and dynamics. Ann Rev Neurosci 12: 157–183.

Basmajian JV, De Luca CJ (1985) Muscles Alive, 5th ed. Williams & Wilkins, Baltimore.

Baud-Bovy G, Soechting JF (2001) Two virtual fingers in the control of the tripod grasp. J Neurophysiol 86: 604–615.

Baud-Bovy G, Soechting JF (2002) Factors influencing variability in load forces in a tripod grasp. Exp Brain Res 143: 57–66.

Bemben MG. Age-related alterations in muscular endurance. Sports Med 25: 259–269, 1998.

Bernstein NA (1947) On the Construction of Movements. Moscow: Medgiz (In Russian).

Bernstein NA (1967) The co-ordination and regulation of movements. Oxford: Pergamon Press.

Bickerton LE, Agur AM, Ashby P (1997) Flexor digitorum superficialis: locations of individual muscle bellies for botulinum toxin injections. Muscle Nerve 20: 1041–1043.

Bigland-Ritchie B, Johansson R, Lippold OC, Woods JJ (1983) Contractile speed and EMG changes during fatigue of sustained maximal voluntary contractions. J Neurophysiol 50: 313–324.

Bigland-Ritchie BR, Dawson NJ, Johansson RS, Lippold OC (1986) Reflex origin for the slowing of motoneurone firing rates in fatigue of human voluntary contractions. J Physiol 379: 451–459.

Binder-Macleod SA, McDermond LR (1992) Changes in the force-frequency relationship of the human quadriceps femoris muscle following electrically and voluntarily induced fatigue. Phys Ther 72: 95–104.

Boatright JR. Kiebzak GM. O'Neil DM. Peindl RD. Measurement of thumb abduction strength: normative data and a comparison with grip and pinch strength. J Hand Surg (Amer). 22: 843–848, 1997.

Burnett RA, Laidlaw DH, Enoka RM. Coactivation of the antagonist muscle does nto covary with steadiness in old adults. J Appl Physiol 89: 61–71, 2000.

Burstedt MK, Edin BB, Johansson RS (1997) Coordination of fingertip forces during human manipulation can emerge from independent neural networks controlling each engaged digit. Exp Brain Res 117: 67–79.

Burstedt MK, Flanagan JR, Johansson RS (1999) Control of grasp stability in humans under different frictional conditions during multidigit manipulation. J Neurophysiol 82, 2393–2405.

Campbell MJ, McComas AJ, Petito F (1973) Physiological changes in aging muscles. J Neurol Neurosurg Psychiat 36: 174–182.

Cavanaugh JT, Shinberg M, Ray L, Shipp KM, Kuchibhatla M, Schenkman M (1999) Kinematic characterization of standing reach: comparison of younger vs. older subjects. Clin Biomech 1999 14: 271–279.

Chao EY, An KN (1978) Graphical interpretation of the solution to the redundant problem in biomechanics. J Biomech Eng 100: 159–167.

Chao EY, Opgrande JD, Axmear FE (1976) Three dimensional force analysis of finger joints in selected isometric hand function. J Biomech 19: 387–396.

Chaput S, Proteau L (1996a) Aging and motor control. J Gerontol. Ser B, Psychol Sci & Social Sci. 51: 346–355.

Chaput S, Proteau L (1996b) Modifications with aging in the role played by vision and proprioception for movement control. Exp Aging Res 22: 1–21.

Christ CB, Boileau RA, Slaughtr MH, Stillman RJ, Cameron JA, Massey BH (1992).

Maximal voluntary isometric force production characteristics of six muscle groups in women aged 25 to 74 years. Am J Human Biol 4: 537–545.

Cole KJ (1991) Grasp force control in older adults. J Mot Behav 23: 251–258.

Cole KJ, Rotella DL, Harper JG (1998) Tactile impairements cannot explain the effect of age on a grasp and lift. Exp Brain Res 121: 263–269.

Cole KJ, Rotella DL, Harper JG. Mechanisms for age-related changes of fingertip forces during precision gripping and lifting in adults. J Neurosci 19: 3238–3247, 1999.

Cole KJ, Abbs JH (1986) Coordination of three-joint digit movements for rapid finger-thumb grasp. J Neurophysiol 55: 1407–1423.

Cole KJ, Abbs JH (1987) Kinematic and electromyographic responses to perturbation of a rapid grasp. J Neurophysiol 57: 1498–1510.

Cole KJ, Johansson RS (1993) Friction at the digit-object interface scales the sensorimotor transformation for grip responses to pulling loads. Exp Brain Res 95: 523–532.

Cole KJ, Rotella DL (2002) Old age impairs the use of arbitrary visual cues for predictive control of fingertip forces during grasp. Exp Brain Res 143: 35–41.

Connelly DM, Rice CL, Roos MR, Vandervoort AA (1999) Motor unit firing rates and contractile properties in tibialis anterior of young and old men. J Appl Physiol 87: 843–852.

Contreras-Vidal JL, Teulings HL, Stelmach GE (1998) Elderly subjects are impaired in spatial coordination in fine motor control. Acta Psychol (Amst) 100: 25–35.

Cooke JD, Brown SH, Cunningham DA (1989) Kinematics of arm movements in elderly humans. Neurol Aging 10: 159–165.

Cooper S, Eccles JC (1930) The isometric responses of mammalian muscles. J Physiol 69: 377–385.

Danion F, Latash ML, Li ZM, Zatsiorsky VM (2000) The effect of fatigue on multi-finger coordination in force production tasks. J Physiol 523: 423–532.

Danion F, Latash ML, Li Z-M, Zatsiorsky VM (2001) The effect of a fatiguing exercise by the index finger on single- and multi-finger force production tasks. Exp Brain Res 138: 322–329.

Danion F, Schöner G, Latash ML, Li S, Scholz JP, Zatsiorsky VM (2003) A force mode hypothesis for finger interaction during multi-finger force production tasks. Biol Cybern 88: 91–98.

Darling WG, Cooke JD, Brown SH (1989) Control of simple arm movements in elderly humans Neurobiol Aging 10: 149–157.

Darling WG, Cole KJ, Miller GF (1994) Coordination of index finger movements. J Biomech 27: 479–491.

Denny Brown DE (1966) The cerebral control of movement. Liverpool Univ. Press, UK.

DeVita P, Hortobagyi T (2000) Age causes a redistribution of joint torques and powers during gait. J Appl Physiol 88: 1804–1811.

Doherty TJ. Brown WF (1997) Age-related changes in the twitch contractile properties of human thenar motor units. J Appl Physiol 82: 93–101.

Duchateau J, Hainaut K (1990) Effects of immobilization on contractile properties, recruitment and firing rates of human motor units. J Physiol 422: 55–65.

Edin BB, Westling G, Johansson RS (1992) Independent control of human finger-tip forces at individual digits during precision lifting. J Physiol 450: 547–564.

Eliasson AC, Forssberg H, Ikuta K, Apel I, Westling G, Johansson, R. (1995) Development of human precision grip. V. Anticipatory and triggered grip actions during sudden loading. Exp Brain Res 106: 425–433.

Enoka RM, Christou EA, Hunter SK, Kornatz KW, Semmler JG, Taylor AM, Tracy BL (2003) Mechanisms that contribute to differences in motor performance between young an old adults. J Electromyogr Kinesiol 13: 1–12.

Era P, Lyyra AL, Viitasalo JT, Heikkinen E (1992) Determinants of isometric muscle strength in men of different ages. Eur J Appl Physiol Occup Physiol 64: 84–91.

Flanagan JR, Wing AM (1995) The stability of precision grip forces during cyclic arm movements with a hand-held load. Exp Brain Res 105: 455–464.

Flanagan JR., Burstedt MK, Johansson RS (1999) Control of fingertip forces in multidigit manipulation. J Neurophysiol 81: 1706–1717.

Fleckenstein JL, Watumull D, Bertocci LA, Parkey RW, Peshock RM. Finger-specific flexor recruitment in humans: depiction by exercise-enhanced MRI. J Appl Physiol 1992; 72: 1974–1977.

Francis KL, Spirduso WW (2000) Age differences in the expression of manual asymmetry. Exp Aging Res 26: 169–180.

Galganski ME, Fuglevand AJ and Enoka RM (1993) Reduced control of motor output in a human hand muscle of elderly subjects during submaximal contractions. J Neurophysiol 69: 2108–2115.

Gelfand IM, Tsetlin ML (1966) On mathematical modeling of the mechanisms of the central nervous system. In: Gelfand IM, Gurfinkel VS, Fomin SV, Tsetlin ML (Eds.) Models of the structural-functional organization of certain biological systems. , pp. 9–26, Nauka: Moscow (in Russian, a translation is available in 1971 edition by MIT Press: Cambridge MA.)

Giampaoli S, Ferrucci L, Cecchi F, Lo Noce C, Poce A, Santaquilani A, Vescio MF, Menotti A (1999) Hand-grip strength predicts insident disability in non-disabled older men. Age Aging 28: 283–288.

Gilles MA, Wing AM (2003) Age-related changes in grip force and dynamics of hand movement. J Mot Behav 35: 79–85.

Grabiner MD, Enoka RM (1995) Changes in movement capabilities with aging. Exer Sport Sci Rev 23: 65–104.

Grimby G, Danneskold-Samsoe B, Hvid K, Saltin B (1982) Morphology and enzymatic capacity in arm and leg muscles in 78–81 year old men and women. Acta physiol scand, 115: 125–134.

Hackel ME, Wolfe GA, Bang SM, Canfield JS (1992) Changes in hand function in the aging adult as determined by the Jebsen Test of Hand Function. Phys Ther 72: 373–377.

Hager-Ross C, Cole KJ, Johansson RS (1996) Grip-force responses to unanticipated object loading: load direction reveals body- and gravity-referenced intrinsic task variables. Exp Brain Res 110: 142–150.

Hakkinen K, Newton RU, Gordon SE, McCormick M, Volek JS, Nindl BC, Gotshalk LA, Campbell WW, Evans WJ, Hakkinen A, Humphries BJ, Kraemer WJ (1998) Changes in muscle morphology, electromyographic activity, and force production characteristics during progressive strength training in young and older men. J Gerontol. Ser A, Biol Sci Med Sci 53: B415–423.

Harding DC, Brandt KD, Hillberry BM (1993) Finger joint force minimization in pianists using optimization techniques. J Biomech 26: 1403–1412.

Hermsdorfer J, Marquardt C, Philipp J, Zierdt A, Nowak D, Glasauer S, Mai N (1999) Grip forces exerted against stationary held objects during gravity changes. Exp Brain Res. 126: 205–214.

Hughes S, Gibbs J, Dunlop D, Edelman P, Singer R, Chang RW (1997) Predictors of decline in manual performance of older adults. J Am Geriatr Soc 45: 905–910.

Iberall T (1997) Human prehension and dexterous robot hands. International J Robot Res 16: 285–299.

Ikeda ER, Schenkman ML, Riley PO, Hodge WA (1991) Influence of age on dynamics of rising from a chair. Phys Ther 71: 473–481.

Jeneson JA, Nelson SJ, Vigneron DB, Taylor JS, Murphy-Boesch J, Brown TR (1992) Two-dimensional 31P-chemical shift imaging of intramuscular heterogeneity in exercising human forearm muscle. Am J Physiol 263: C357–364.

Johansson RS (1996) Sensory and memory information in the control of dexterous manipulation. In: Lacquaniti F, Viviani P (Eds.) Neural Bases of Motor Behavior, pp. 67–79, Kluwer Acad.Publ.: Amsterdam.

Kamen G, Sison SV, Du CC, Patten C (1995) Motor unit discharge behavior in older adults during maximal-effort contractions. J Appl Physiol 79: 1908–1913.

Kenshalo DR (1979) Sensory function of the skin. Plenum Press: New York.

Kernell D, Eerbeek O, Verhey BA (1983) Relation between isometric force and stimulus rate in cat's hindlimb motor units of different twitch contraction time. Exp Brain Res 50: 220–227.

Kilbreath SL, Gandevia SC (1994) Limited independent flexion of the thumb and fingers in human subjects. J Physiol 479: 487–497.

Kinoshita H, Francis PR (1996) A comparison of prehension force control in young and elderly individuals. Eur J Appl Physiol 74: 450–460.

Kinoshita H, Kawai S, Ikuta K (1995) Contributions and co-ordination of individual fingers in multiple finger prehension. Ergonomics 38: 1212–1230.

Kirkendall DT, Garrett WE Jr (1998) The effects of aging and training on skeletal muscle. Amer J Sports Med 26: 598–602.

Laidlaw DH, Hunter SK, Enoka RM (2002) Nonuniform activation of the agonist muscle does not covary with index finger acceleration in old adults. J Appl Physiol 93: 1400–1410.

Landsmeer JM, Long C (1965) The mechanism of finger control, based on electromyograms and location analysis. Acta Anat 60: 330–347.

Larsson L, Ansved T (1995) Effects of ageing on the motor unit. Prog Neurobiol 45: 397–458.

Latash ML (1993) Control of Human Movement. Human Kinetics: Urbana, IL.

Latash ML (1996) How does our brain make its choices? In: Latash ML, Turvey MT (Eds.) Dexterity and Its Development, pp. 277–304, Erlbaum Publ.: Mahwah, NJ.

Latash ML, Li ZM, Zatsiorsky (1998) A principle of error compensation studied within a task of force production by a redundant set of fingers. Exp Brain Res 122:131–138.

Latash ML, Scholz JF, Danion F, Schöner G (2001) Structure of motor variability in marginally redundant multi-finger force production tasks. Exp Brain Res 141: 153–165.

Latash ML, Scholz JP, Schöner G (2002a) Motor control strategies revealed in the structure of motor variability. Exer Sport Sci Rev 30: 26–31.

Latash ML, Li S, Danion F, Zatsiorsky VM (2002b) Central mechanisms of finger interaction during one- and two-hand force production at distal and proximal phalanges. Brain Res 924: 198–208.

Latash ML, Scholz JF, Danion F, Schöner G (2002c) Finger coordination during discrete and oscillatory force production tasks. Exp Brain Res 146: 412–432.

Latash ML, Shim JK, Zatsiorsky VM (2004) Is there a timing synergy during multi-finger production of quick force pulses. Exp Brain Res 159: 65–71.

Leijnse JN, Snijders CJ, Bonte JE, Landsmeer JM, Kalker JJ, Van der Meulen JC, Sonneveld GJ, Hovius SE (1993) The hand of the musician: the kinematics of the bidigital finger system with anatomical restrictions. J Biomech 26: 1169–1179.

Li S, Danion F, Latash ML, Li Z-M, Zatsiorsky VM (2000) Characteristics of finger force production during one- and two-hand tasks. Hum Move Sci 19: 897–924.

Li S, Latash ML, Zatsiorsky VM (2003) Finger interaction during multi-finger tasks involving finger addition and removal. Exp Brain Res 150: 230–236.

Li Z-M, Latash ML, Zatsiorsky VM (1998) Force sharing among fingers as a model of the redundancy problem. Exp Brain Res. 119: 276–286.

Li Z-M, Zatsiorsky VM, Latash ML (2000) Contribution of the extrinsic and intrinsic hand muscles to the moments in finger joints. Clin Biomech 15: 203–211.

Long C (1965) Intrinsic-extrinsic muscle control of the fingers. J Bone Joint Surg Am 50: 973–984.

McDonagh MJ, White MJ, Davies CT (1984) Differential effects of ageing on the mechanical properties of human arm and leg muscles. Gerontology, 30: 49–54.

MacKenzie CL, Iberall T (1994) The Grasping Hand. North Holland, Amsterdam.

Miller AE, MacDougall JD, Tarnopolsky MA, Sale DG (1993) Gender differences in strength and muscle fiber characteristics. Eur J Appl Physiol Occup Physiol 66: 254–262.

Mussa Ivaldi FA, Morasso P, Zaccaria R (1989) Kinematic networks. A distributed model for representing and regularizing motor redundancy. Biol Cybern 60: 1–16

Nakao M, Inoue Y, Murkami H (1989) Aging process of leg muscle endurance in males and females. Eur J Appl Physiol, 59: 209–214.

Ohtsuki T (1981) Inhibition of individual fingers during grip strength exertion. Ergonomics 24: 21–36.

Owings TM. Grabiner MD (1998) Normally aging older adults demonstrate the bilateral deficit during ramp and hold contractions. J Gerontol Ser A, Biol Sci & Med Sci. 53: B425–429.

Penfield W, Boldrey E (1937) Somatic motor and sensory representation in the cerebral cortex of man as studied by electrical stimulation. Brain 37: 389–443.

Penfield W, Rasmussen T (1950) The Cerebral Cortex of Man, New York: MacMillan.

Pratt J, Chasteen AL, Abrams RA (1994) Rapid aimed limb movements: Age differences and practice effects in component submovements. Psychol Aging 9: 325–334.

Prilutsky BI, Zatsiorsky VM (2002) Optimization-based models of muscle coordination. Exer Sport Sci Rev 30: 32–38.

Ranganathan VK, Siemionow V, Sahgal V, Liu JZ, Yue GH (2001) Skilled finger movements exercise improves hand function. J Gerontol A Biol Sci Med Sci 56: M518–M522.

Rantanen T, Guralnik JM, Foley D, Masaki K, Leveille S, Curb JD, White L (1999) Midlife hand grip strength as a predictor of old age disability. JAMA 281: 558–560.

Rosenbaum DA, Loukopoulos LD, Meulenbroek RGM, Vaughan J, Engelbrecht SE (1995) Planning reaches by evaluating stored postures. Psychol Rev 102: 28–67.

Rouiller EM (1996) Multiple hand representations in the motor cortical areas. In: Wing AM, Haggard P and Flanagan JR (Eds.) Hand and Brain. The Neurophysiology and Psychology of Hand Movements. pp. 99–124. Academic Press, San Diego, CA.

Santello M, Soechting JF (2000) Force synergies for multifingered grasping. Exp Brain Res 133: 457–67.

Schieber MH (1991) Individuated finger movements of rhesus monkeys: a means of quantifying the independence of the digits. J Neurophysiol 65: 1381–1391.

Schieber MH (2001) Constraints on somatotópic organization in the primary motor cortex. J Neurophysiol 86: 2125–2143.

Scholz JP, Schöner G (1999). The uncontrolled manifold concept: Identifying control variables for a functional task. Exp Brain Res 126, 289–306.

Scholz JP, Danion F, Latash ML, Schöner G (2002) Understanding finger coordination through analysis of the structure of force variability. Biol Cybern 86: 29–39.

Seidler-Dobrin RD, He J, Stelmach GE (1998) Coactivation to reduce variability in the elderly. Motor Control 2: 314–330.

Semmler JG, Steege JW, Kornatz KW, Enoka RM (2000) Motor-unit synchronization is not responsible for larger motor-unit forces in old adults. J Neurophysiol 84: 358–366.

Serlin DM, Schieber MH (1993) Morphologic regions of the multitendoned extrinsic finger muscles in the monkey forearm. Acta Anat 146: 255–266.

Shim JK, Latash ML, Zatsiorsky VM (2003a) Prehension synergies: Trial-to-trial variability and hierarchical organization of stable performance. Exp Brain Res 152: 173–184.

Shim JK, Latash ML, Zatsiorsky VM (2003b) The central nervous system needs time to organize task-specific covariation of finger forces. Neurosci Lett 353: 72–74.

Shim J.K., Lay B., Zatsiorsky V.M., Latash M.L. (2004) Age-related changes in finger coordination in static prehension tasks. J Appl Physiol 97: 213–224.

Shinohara M, Li S, Kang N, Zatsiorsky VM, Latash ML (2003a) Effects of age and gender on finger coordination in maximal contractions and submaximal force matching tasks. J Appl Physiol 94: 259–270.

Shinohara M, Latash ML, Zatsiorsky VM (2003b) Age effects on force production by the intrinsic and extrinsic hand muscles and finger interaction during maximal contraction tasks. J Appl Physiol 95: 1361–1369.

Shinohara M, Scholz JP, Zatsiorsky VM, Latash ML (2004) Finger interaction during accurate multi-finger force production tasks in young and elderly persons. Exp Brain Res 156: 282–292.

Smith RJ (1974) Balance and kinetics of the fingers under normal and pathological conditions. Clin Orthop 104: 92–111.

Spiegel KM. Stratton J. Burke JR. Glendinning DS. Enoka RM (1996) The influence of age on the assessment of motor unit activation in a human hand muscle. Exp Physiol 81: 805–819.

Stelmach GE, Goggin NL, Garcia-Colera A (1987) Movement specification time with age. Exp Aging Res 13: 39–46.

Stelmach GE, Goggin NL, Amrheim PC (1988) Aging and the restructuring of precued movements. Psychol Aging 3: 151–157.

Thomas CK, Bigland-Richie B, Johansson RS (1991) Force-frequency relationships of human thenar motor units. J Neurophysiol 65: 1509–1516.

Tuite DJ. Renstrom PA. O'Brien M (1997) The aging tendon. Scand J Med Sci Sp 7: 72–77.

Turvey MT (1990) Coordination. Amer Psychol 45: 938–953.

Vaillancourt DE, Larsson L, Newell KM (2003) Effects of aging on force variability, single motor unit discharge patterns, and the structure of 10, 20, and 40 Hz EMG activity. Neurobiol Aging 24: 25–35.

Vaillancourt DE, Newell KM (2003) Aging and the time and frequency structure of force output variability. J Appl Physiol 94: 903–912.

Viitasalo JT, Era P, Leskinen AL, Heikkinen E (1985) Muscular strength profiles and anthropometry in random samples of men aged 31–35, 51–55 and 71–75 years. Ergonomics 28: 1563–1574.

Welford AT (1984) Psychomotor performance. Ann Rev Gerontol Geriatr, 4: 237–273.

Weeks DL, Sherwood DE, Noteboom JT (2002) Anticipatory modulation of precision grip force with variations in limb velocity of a curvilinear movement. J Mot Behav 34: 59–66.

Winegard KJ. Hicks AL. Vandervoort AA (1997) An evaluation of the length-tension relationship in elderly human plantarflexor muscles. J Gerontol Ser A, Biol Sci Med Sci. 52: B337–343.

Woollacott M, Inglin B, Manchester D (1988) Response preparation and posture control. Neuromuscular changes in the older adult. Ann New York Acad Sci 515: 42–53.

Zatsiorsky VM, Li Z-M, Latash ML (1998) Coordinated force production in multi-finger tasks. Finger interaction, enslaving effects, and neural network modeling. Biol Cybern 79: 139–150.

Zatsiorsky VM, Gregory RW, Latash ML (2002a) Force and torque production in static multi-finger prehension: Biomechanics and Control. Part I. Biomechanics. Biol Cybern 87: 50–57.

Zatsiorsky VM, Gregory RW, Latash ML (2002b) Force and torque production in static multi-finger prehension: Biomechanics and Control. Part II. Control. Biol Cybern 87: 40–49.

Zatsiorsky VM, Gao F, Latash ML (2003) Prehension synergies: Effects of object geometry and prescribed torques. Exp Brain Res 148: 77–87, 2003.

Zimmerman SD, McCormick RJ, Vadlamudi RK and Thomas DP (1993) Age and training alter collagen characteristics in fast- and slow- twitch rat limb muscle. J Appl Physiol 75: 1670–1674.

AUTHOR INDEX

SUBJECT INDEX